A NATIONAL TRUST GUIDE

GREAT AMERICAN BRIDGES AND DAMS

A NATIONAL TRUST GUIDE

GREAT ☆ AMERICAN ☆ BRIDGES AND DAMS

DONALD C. JACKSON

Foreword by David McCullough

GREAT AMERICAN PLACES SERIES
PRESERVATION PRESS

John Wiley & Sons, Inc.
New York   Chichester   Brisbane   Toronto   Singapore

Printed in the United States of America
10 9 8

Library of Congress Cataloging in Publication Data

Jackson, Donald C. (Donald Conrad), 1953–
    Great American bridges and dams / Donald C. Jackson
        p.   cm.
    At head of title: A National Trust guide.
    Bibliography: p.
    Includes index.
    ISBN 0-471-14385-5
    1. Bridges — United States.   2. Dams — United States.
I. National Trust for Historic Preservation in the United
States.   II. Title.
TG23.J33   1988
624'.2'0973—dc19   87-22309

Designed by Meadows & Wiser, Washington, D.C.

Composed in Vladim by General Typographers, Inc.,
Washington, D.C.

Front cover: View of the spillway of the New Croton Dam,
Croton-on-Hudson, N.Y., c. 1907. (Postcard courtesy of
Donald C. Jackson)
Back cover: Waco Suspension Bridge, Waco, Tex.,
c. 1910. (Postcard courtesy of Donald C. Jackson)

CONTENTS

■ ■ ■ ■ ■ THE BRIDGES AND DAMS ■ ■ ■ ■ ■ ■

BOULDER DAM, COLORADO

GEORGE WESTINGHOUSE MEMORIAL BRIDGE AND PLANTS ON LINCOLN HIGHWAY, EAST PITTSBURGH, PA.

███████████████████████

FOREWORD

Bridges over great rivers and mountain ravines, bridges to carry a nation — and the ideal of the good society — forward, across the "untamed" continent . . . bridges for growth, profit, progress, for pride in good old Brooklyn or good old Kansas City or the Golden Gate . . . we have loved them all. It was American, everybody said, to be so good at things mechanical and to dream large. It was a European, John A. Roebling, who perfected the suspension bridge, the most elegant bridge form, but he did it *here*—that was the point, we knew. "Go find your destiny in all that space" was the admonition of his mentor Hegel in Berlin, when Roebling could only imagine America. Bridges into the future — for Americans it has always been the future that counted. "The shapes arise!" proclaimed Walt Whitman as Roebling's masterpiece got under way at Brooklyn after the Civil War.

It takes energy and ingenuity and a degree of wealth to build even a modest bridge. But it takes also the desire to get from here to there, and nowhere was the desire so great as with the American people. Foolish things were said about conquering nature; they still are. The builders succeeded when they worked with the forces of nature, not against, as every first-rank civil engineer understands and as so many bridges attest. The best of them stir us as music does.

Bridges define our landscape. They figure in the vision of our painters (Thomas Eakins, Joseph Stella, Charles Sheeler, John Kane) and our writers (Willa Cather, Hart Crane, Arthur Miller). "Before the war I built bridges," says the soldier-hero, Jimmy Stewart, to the lovely Chinese woman, Lisa Lu, in the movie *The Mountain Road*, and we know at once he is a good man. That is how we feel about bridges.

The first bridge I took time to look at stood on a window sill in the sunshine in a schoolroom in Pittsburgh. I was 9 or 10. It was Miss Scheltz's room. She taught science with enthusiasm and imagination. The bridge was a long, bowstring truss made of matchsticks, the project of a boy from one of the older classes. I wish I knew his name. What he had built seemed a miracle to me, truly, unforgettably beautiful. Afterward I began seeing the real thing. Pittsburgh, we heard, had more bridges than Paris. At Smithfield Street downtown, close to where my father worked, an extraordinary silver bridge carried automobile traffic and streetcars over the brown Monongahela. I've since learned it is a lenticular truss, a very rare species, and one of the most striking bridges in the country. It still makes me feel good about my hometown. Its designer, Gustav Lindenthal, ought to be remembered and honored as we honor our best-known architects — just as the names Charles Ellet, James B. Eads, George S. Morison, Othmar H. Ammann, Ralph

Gustav Lindenthal's Smithfield Street Bridge, Pittsburgh. Lindenthal's original portal structures were replaced in 1915 with this cast-steel design.

John A. Roebling's Brooklyn Bridge. The Manhattan Bridge is in the background.

Modjeski, David B. Steinman, so many who figure in these pages, should rank among our most valued Americans.

This book is, of course, a guide to dams as well, a subject I know less about. It is well they are here, wonderful to have the greatest of them so cataloged, as structural triumphs like bridges.

If what we build says as much about us as anything, and I believe it does, then this handsome, valuable volume, so long needed, is also a portrait of America.

David McCullough

PREFACE

T o combine structures as different as bridges and dams into a single guidebook might at first appear incongruous. Bridges are designed to carry people, vehicles and material products over both natural and man-made barriers, while dams are constructed to impound or divert water along rivers and streams. Little seems to connect the two. But bridges and dams are bound by a common heritage. Both are among the most visible and most important manifestations of civil engineering in our environment, and both are essential components of the public works foundation that supports America's transportation, electric power, agricultural and water supply systems.

Civil engineering encompasses a wide range of activities and is essential to endeavors such as skyscraper design, airport construction, highway building and sewage disposal, to name but a few. Many of the pleasures of late 20th-century life often taken for granted are dependent on the work of civil engineers, work that is frequently hidden or obscured from the public's view. Bridges and dams are major exceptions to this phenomenon — they often stand as key symbols of civic development. For this reason, these two seemingly diverse types of structures have been combined here.

Bridges and dams, as with all technological artifacts, are significant only because of what they do, not what they are. They are not large expensive sculptures erected primarily for aesthetic purposes. They are important because of their usefulness within larger systems that support social, cultural and economic development. To be sure, their visual appearance can be a vital aspect of their design and should not be discounted. But, at the root, bridges and dams are built to serve practical, utilitarian functions, and usefulness is the essence of their existence.

Consequently, the descriptions of many of the sites listed in this book go beyond simple data related to their dates and dimensions. The bridges and dams have been placed in historical contexts that illuminate their technological origins, the nature of their operation or their role in the local region's socioeconomic development. These analyses are by no means comprehensive, but they are designed to demonstrate the significance of these structures in America's history.

Donald C. Jackson

ACKNOWLEDGMENTS

A work of this magnitude could not have been completed without the help of many people, but two deserve special recognition. First and foremost, I extend heartfelt appreciation to my mother, Jessie Tufts Jackson, who encouraged me throughout the project. In particular, she provided superb editorial assistance and allowed me to concentrate on substantial issues of research and writing. Second, I would like to thank Mrs. Hobart Bosworth of Littlerock, Calif. As president of the Littlerock Creek Irrigation District and as a scholar with uncanny insight into the meaning of history, she helped open my eyes to the special significance of water in the development of the American West. Her tireless work in the civic improvement of northern Los Angeles County demonstrates that people, history and preservation can make a difference in the quality of a community's life. More than just a celebration of construction, this book strives to interest people in the *value* of our public works heritage. Mrs. Bosworth first impressed on me the importance of such values in regional social and economic development, and for this I am deeply grateful.

Several scholars knowledgeable in engineering history provided valuable comments and recommendations after reviewing portions of the manuscript. Their insights have contributed substantially to the text. The author, of course, retains responsibility for any errors. Many thanks to William P. Chamberlin, Clayton Fraser, Charles K. Hyde, David Simmons and John Snyder.

Others who assisted include Herbert F. Hands, American Society of Civil Engineers; Bruce Clouette and Matt Roth, Historic Resources Consultants; Susan Munkres, Water Resources Center Archives; David Introcaso, Salt River Project; Elaine Howard, Montana Power Company; Martha Carver, Tennessee Department of Transportation; George Hauck, University of Missouri at Kansas City; Mary Ison and Ford Peatross, Prints and Photographs Division, Library of Congress; David Shayt, Robert M. Vogel and William Worthington, National Museum of American History, Smithsonian Institution; Marty Reuss, Office of History, and William F. Willingham, Portland District, U.S. Army Corps of Engineers; and Ian Spatz of the National Trust.

Present and past staff members of the National Park Service's Historic American Engineering Record have assisted me on bridge and dam projects during the past 13 years. They include Richard Anderson, Marjorie Baer, T. Allan Comp, Eric Delony, Gray Fitzsimons, Douglas Griffin, Arnold David Jones, Robert Kapsch, Larry Lankton, Jet Lowe and Jean Yearby. In addition, Alison Hoagland of the Park Service's Historic American Buildings Survey, Mary Farrell of the National Register of Historic Places, Ann Huston of the Western Regional Office and Greg Kendrick of the Rocky Mountain Regional Office aided in the book's preparation.

The following individuals from state historic preservation offices, state historical societies and other private and public agencies provided essential photographs and guidance:

Alabama: Alabama Historical Commission. Arkansas: Michael Swanda. California: Steve Mikesell, Will Rivera, Anna Rodriques and Don Westphal. Colorado: Kaaren K. Patterson and Leslie E. Wildesen. Connecticut: Dawn Maddox; Cora Murray and Dave Poirier. Delaware: Alice H. Guerrant. District of Columbia: Mary Kaye Freedman and Donald B. Myer. Florida: Cookie O'Brien and William Thurston. Georgia: Tom and Larry French, Helen Stacy and Ken Thomas. Guam: Edward Pangelinian. Idaho: Elizabeth Jacox; David D. Le Pard and Merle W. Wells. Illinois: Julia Hertenstein and Donna Pruitt. Indiana: James L. Cooper and Joan E. Hostetler. Iowa: Marita Moir. Kansas: Larry Jochims. Louisiana: Judith H. Bonner. Maine: Kirk F. Mohney. Maryland: Orlando Rideout IV. Massachusetts: Leanne Del Vecchio, Joe Russell and Margaret Twomey. Michigan: Robert Christianson. Minnesota: Dennis Gimmestad and Dona Sieden. Mississippi: Susan M. Enzweiler. Montana: Pat Bick. Nebraska: Marty Miller and James E. Potter. Nevada: Ronald M. James. New Hampshire: Dale Ford. New Jersey: Charles Cummings and Terry Karschner. New Mexico: Mary Ann Anders. New York: Marion Bernstein and Larry Gobrecht. North Carolina: Michael T. Southern. North Dakota: Forrest Daniel and James E. Sperry. Ohio: Tauni Graham. Oklahoma: Loweta Chesser. Oregon: Dwight A. Smith. Pennsylvania: Joseph J. Ellam and Tobi Gilson. South Carolina: Nancy Pittenger. South Dakota: John E. Rau. Tennessee: James F. Jones and Claudette Stager. Texas: Chris Clymer, Casey Greene, Jane Kenmore and William H. Richter. Utah: Utah State Historical Society. Vermont: Gina Campoli and Candice J. Deininger. Virginia: Beth Hoge. West Virginia: Rodney S. Collins. Wisconsin: Paul Lusignan. Wyoming: Richard Collier.

My special thanks to Pamela Haag of Swarthmore College and Elizabeth Hughes of Georgetown University, whose service as Preservation Press interns greatly aided in gathering photographs from a wide variety of sources. My appreciation also to Diane Maddex, Janet Walker and Michelle LaLumia of the Preservation Press who, together with Gretchen Smith, oversaw all phases of publication from development and editing through photo research and production.

Thanks also to David Cobb of Phillips Academy, who gave me confidence in my ability to write in the English language. And finally, I acknowledge my debt to the late Samuel Carpenter, professor of engineering at Swarthmore College, who first encouraged me to study the history of civil engineering. I think he would be proud that one of his students wrote this book.

BRIDGES: SPANNING THE NATION

From pre-history transportation has played a key role in the growth of civilization. In early societies, rivers formed natural barriers to the movements of tribal groups, and, without doubt, the first bridges were fallen trees placed across rushing streams. As cultures developed, primitive peoples expanded upon these ready-made bridges in their search for improved trade and transportation, erecting slabs of stone or simple log bridges to span slow-moving rivers. As communities grew even more complex, especially during the Chinese and Roman empires, bridge construction evolved into a major form of engineering. Roman engineers, in particular, experimented with and perfected huge stone arch structures that were vitally important in supporting the growth of cities and binding together the far-flung Roman Empire.

With the decline of Rome, European bridge building languished, and during the Middle Ages only a few new bridges were erected. Bridge engineering, however, never died out completely, and the technology of stone arch construction, exemplified in such spans as the Avignon Bridge in France (1187) and the original London Bridge (1209), survived in its basic form. In the 15th and 16th centuries the Renaissance, which inspired an interest in the study of extant Roman artifacts, fostered a revival of bridge building using technologies based on both stone arch and timber construction.

When Europeans began settling in eastern North America in the 17th century, little effort was made to erect permanent crossings over waterways. Settlers usually forded small streams, but sometimes they constructed wooden pile bridges that resembled small, temporary piers or wharves. These wooden structures were susceptible to rotting and could be washed out easily by spring floods. Despite their impermanence, they represented logical engineering solutions to the problem at hand: they did not require extensive amounts of labor to build, they used local materials, and they could be quickly rebuilt if destroyed. They also required only rudimentary design and construction skills. On larger rivers, boats and ferries were used to carry people and produce from one shore to another. Although infrequently encountered as part of America's modern highway system, ferries were a common means of crossing rivers until well into the 20th century. They did not require large investments of capital and were particularly well suited to rural areas where the amount of highway traffic did not justify construction of a permanent crossing.

By the end of the 18th century, many people began to recognize the importance of building a permanent, reliable system of roads to bind together the newly formed United States of America. For example, delays in travel, caused by the poor condition of highways and river crossings, impeded the arrival of delegates to the

Opposite: One of the towers supporting the San Francisco–Oakland Bay Bridge (1935), the longest high-level bridge in the world.

Right: Log across a stream — the earliest type of bridge — in a park setting, Norwich, England, c. 1900. Below: Remains of the Avignon Bridge in France, built by 12th-century Benedictine monks.

Typical wooden pile bridge near East Machias, Maine, c. 1910.

Constitutional Convention in Philadelphia in the summer of 1787. The improvement of highway travel thus became the first important function of bridges in the United States.

In the early 19th century canal construction flourished as a means of improving inland transportation, and the network of canals created a new demand for bridge construction. Many bridges were built to provide access over canals, but numerous structures, known as aqueducts, also were built to carry the canals themselves over streams and other natural barriers. With the growth of large-scale, urban water supply systems beginning in the mid-19th century, another type of aqueduct developed in America. These aqueducts did not carry water as part of a transportation system; instead, they carried pipelines that supplied water for domestic, agricultural or industrial use.

In the mid-19th century the growth of railroads created a new use for bridges akin to carrying highway traffic. Railroads placed heavier loads on bridges than did horse-drawn wagons, prompting construction of larger and stronger structures. When these railroad bridges

Left: Green River Ferry near Mammoth Cave, Ky., c. 1940. Bridges often replaced ferry services. Below: Tunkhannock Creek Viaduct (1916), a concrete arch bridge built to carry railroad traffic across a wide valley in northeastern Pennsylvania.

(and occasionally even a highway bridge) cross a wide valley, they often are called viaducts, a term derived from Latin. Viaducts are simply long bridges that maintain a constant grade (i.e., they don't have dips or bumps in them) over terrain such as a wide river or valley.

Bridge construction in America during the past 150 years has focused primarily on highways, railroads, canals and water supply systems. This seemingly limited number of functions presumably would have prompted construction of only a few different types of bridges. But as all travelers who have kept their eyes open during even a short highway trip know, the American landscape contains an enormous variety of bridges. In fact, the number of different structural forms built to carry human commerce over natural barriers, even within a relatively small geographical region, is remarkable.

The type of bridge built for a particular crossing depended on many factors, including the site's geological and topographical conditions, the skills and availability of local workers, the price and availability of structural materials, the political and economic influence of regional commercial interests, the visual prominence of the

Kittanning, Pa., c. 1920. A Pennsylvania through truss (right) replaced the 19th-century bowstring arch truss highway bridge — an example of the rapid evolution of design appropriate for a site.

setting, the nature of traffic intended for the span and the previous experience of the design engineer or builder. The influence of each of these factors could vary so much that a design suitable for a given site and function at one time could be inappropriate or ill-advised for the same site 20 years later.

Often an engineering problem has no single best solution. For this reason the history of bridge engineering does not follow a linear path in which one type of technology is used for a while and then rendered universally obsolete by some new technology. To be sure, certain types of bridges ultimately lost their practicality or usefulness as other types were developed. But the story does not consist of even, regular seams stitched together in a neat, orderly pattern. Rather, the seams are often ragged and irregular, challenging the sanity of historians who attempt to analyze bridges in an orderly fashion. Clearly, bridge engineering is not an illogical endeavor practiced with no interest in achieving economically rational results. This does not mean, however, that every bridge builder in a given circumstance will arrive at the same design solution. This fact of life may not make the study of bridge history easy, but it does make the subject infinitely more interesting.

STONE ARCH BRIDGES

With the beginning of permanent bridge construction in America during the late 18th century, one of the first types of structures built was the masonry or stone bridge. The vast majority of these were arch bridges constructed using materials gathered or quarried near the crossing site. Stone arch bridges were built by the Romans for their highways and aqueducts, and traditional construction techniques survived in Europe through the Middle Ages and into the modern period. Stone arch construction was relatively labor intensive, especially for large structures, and this aspect of the technology was not particularly suited to the sparsely populated New World. In 19th-century America stone arch bridges were never as popular as they were in Europe. Nonetheless, engineers and builders still employed them for a variety of functions.

Most stone arch bridges were relatively small-scale

structures built by local masons, but several large examples were built on the National Road, which extended from the Potomac River to the Ohio River and westward, as well as in other parts of the eastern United States. When carefully built, stone arch bridges were capable of carrying heavy loads and they required relatively little maintenance. As a result, they were often used in aqueduct construction, where permanence and long-term stability were important objectives. Railroads also found them attractive because of their strength and rigidity, although in the early to mid-19th century, construction costs often precluded their use. Near the end of the century, stone arch designs experienced a revival as wealthy railroad companies sought to build permanent structures capable of carrying heavy loco-motives and rolling stock. In addition, small-scale stone arch construction for rural highway bridges flourished in isolated areas into the early 20th century, and in the 1930s the technology even became the focus of several Civilian Conservation Corps work projects. On the whole, however, stone arch construction remained sec-ondary in America, especially after iron and steel production boomed in the late 19th century.

Top: Stone arch span (c. 1830), Hillsboro, N.H. Center left: European 19th-century stone arch bridge, reflecting craftsmanship rarely found in American masonry spans. Center right: Pont du Gard, a Roman relic in southern France. Bottom: Late 19th-century stone arch railroad bridge, Springfield, Mass.

WOODEN BRIDGES

Colonial America was blessed with an abundant supply of forests. Wood was readily available almost everywhere and became a vital material in the country's initial economic development. In the 17th century it found use in small-scale, temporary pile bridges across small streams. By the beginning of the 19th century, some of America's more daring builders, such as Timothy Palmer (1751–1821) and Louis Wernwag (1769–1843), started building large-scale wooden arch bridges across major rivers such as the Schuylkill in Philadelphia. However, the most pervasive type of wooden bridge was the simple covered truss bridge, forever immortalized in the romance of American folklore.

Despite popular belief, covered bridges were not invented in America. Large-scale structures of this type were built in continental Europe during the 18th century, and the technology can be traced even to designs developed by the 16th-century Italian architect Andrea Palladio. But, compared to the extensive construction of wooden bridges in America, these European examples seem relatively minor precursors. Before discussing the nature of wooden truss designs, it is worthwhile to affirm that covered bridges were covered to protect the structure from deterioration. Contrary to folk wisdom, they were not designed to prevent horses from becoming frightened or to produce rural "lovers' lanes." At times, wooden bridges were built without protective cladding, but such structures were more susceptible to rotting and required more attention to maintenance.

Truss bridges, whether of wood or metal (or a combination of the two), are characterized by a structural assemblage of many relatively small members joined

Right: Covered bridge (19th-century), Fribourg, Germany, illustrating that covered bridges were not unique to the United States. Below: Another kind of covered bridge in Bondville, Vt., c. 1910. Only the main trusses are covered to protect the structural members.

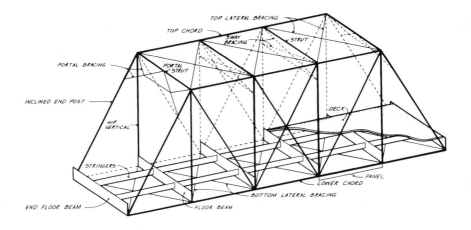

together in a series of triangles that interconnect to form the bridge. One reason early builders and engineers were attracted to truss bridges was the comparative ease of fabricating, hauling and assembling these individual members. The specific type of truss design depends on the arrangement of members in the truss and the nature of the forces they are called on to resist. Truss members are placed either in tension (i.e., forces are acting to pull it apart from either end) or in compression (i.e., forces are acting to push it together from either end). Truss members are either stiff, heavy struts or posts or thin, flexible rods or bars. Stiff struts or posts are capable of withstanding both tension and compression, but thin rods or bars are capable only of withstanding tension. In general, truss members can be distinguished as being in either tension or compression.

In addition to the structural configuration of their

Schematic diagram of a typical truss.

Truss elevations and transverse sections. Top: Through truss. Center: Pony truss. Bottom: Deck truss.

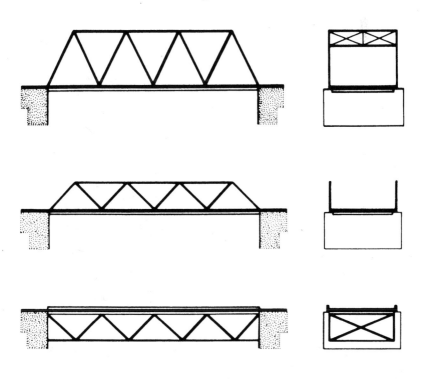

King-post truss, Tawas, Mich., c. 1910.

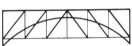

Queen-post truss near Estes Park, Colo., c. 1940.

Interior view of a covered bridge, a Burr arch truss in Montague, Mass., c. 1940.

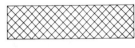

Town lattice truss, Easthampton, Mass., c. 1895.

members, trusses are further categorized according to the location of the traffic deck. Bridges in which the traffic is carried across the top of the truss structure are called deck trusses. If traffic is carried along the bottom chords of the structure, the bridge is called a through truss (because one appears to travel through the truss). And if traffic is carried along the bottom chord but there is no lateral bracing between the top chords of the truss, the bridge is called a pony truss. The origin of this term is unknown, but it probably derives from the fact that most pony trusses are relatively small structures, and a pony is, of course, a small horse.

The earliest wooden trusses were simple structures known generically as king-post trusses. These are short, triangular structures with top-chord compression members and a vertical tension member. At times the basic king-post form could be expanded into a multiple king-post truss that incorporates a series of vertical and diagonal members into its design. The form could also be transformed into another simple design that came to be known as the queen-post truss. This type of truss has two vertical tension members, compared to one in the king-post design.

In the early 19th century a covered-bridge builder from Connecticut, Theodore Burr (1771–1822), took a multiple king-post truss and strengthened it with an auxiliary arch to form the Burr arch truss. Later engineers added arches to other types of trusses to stengthen them, but the Burr arch truss was by far the most commonly built of this type of structure. Strictly speaking, the arch component is not part of the truss proper. It proved extremely useful, however, in reducing the amount of deflection (sag) in the center of the truss, a frequent problem with wooden bridges.

Burr patented his design in 1817, and three years later architect and builder Ithiel Town (1784–1844), also from Connecticut, patented a truss type that also became very popular. Known as the Town lattice truss, this design featured an extensive web, or lattice, of members joined together to form a long, stiff structure. Because of its stiffness, the Town lattice truss was not susceptible to sagging. It also used relatively small members that could easily be transported to a bridge site and assembled by carpenters. Town's truss required extensive drilling of the members to form the holes for the structural connections, but this requirement did not prevent the design from finding wide usage.

Before the Civil War, numerous engineers and builders developed their own special wooden truss designs. Some of these, such as Stephen Long (1784–1864), designed trusses that achieved considerable importance within certain regions. But many other designs were used for only a few projects and constituted little more than engineering curiosities. By the mid-19th century wooden bridges were rapidly being eclipsed in importance by iron structures, although in certain areas (such as Oregon) wood continued to be used for bridges well into the 20th century.

METAL TRUSS BRIDGES

Truss bridge technology was readily adaptable to metal components, and the late 19th century was the heyday of the all-metal truss bridge in America. The change from wooden to metal trusses did not occur abruptly. Rather, it was a gradual transformation that began in the 1840s with the first construction of the Howe truss bridge. Developed by William Howe (1803–52), a native of western Massachusetts and uncle of sewing machine inventor Elias Howe, the Howe truss is a combination design with diagonal wooden compression members and vertical iron tension members. Howe trusses were commonly used by railroads eager to build inexpensive yet relatively strong bridges that used large amounts of cheap wood. Occasionally, Howe truss bridges were built entirely of metal (parts of such a structure dating to the mid-1840s are on display at the Smithsonian Institution's National Museum of American History), but in general the design featured both iron and wood components.

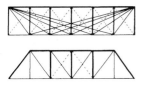

Top: Bollman truss. Above: Pratt truss.

Throughout the late 19th century, engineers developed a wide range of truss types designed primarily to be built of cast iron, wrought iron and, later, steel. Some of these, such as the Bollman truss (developed by Wendel Bollman), the Fink truss (developed by Albert Fink), the Post truss (developed by Simeon S. Post) and the Thacher truss (developed by Edwin Thacher), achieved a substantial degree of success before they faded from popularity. Many others were never much more than a fanciful concoction described in a patent application and used, at best, only a few times. Three truss designs, however, are worthy of special mention: the Pratt truss, the bowstring arch truss and the Warren truss.

Below: Camelback truss. Center: Baltimore trusses. Bottom: Pennsylvania trusses.

The first of these, patented by Thomas (1812–75) and Caleb Pratt in 1844, was a design with vertical compression members and diagonal tension members. The Pratts' patent application illustrated a combination structure with wooden compression members, although the vast majority of Pratt trusses built in the United States were entirely of metal. The Pratt truss achieved enormous popularity because of its strength and straightforward design. It was not a complicated structure that required complex shop work, and it was adaptable to a wide variety of situations.

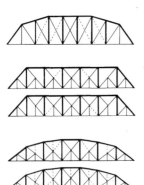

The Pratt truss spawned a variety of related designs: (1) the double-intersection Pratt truss, in which the diagonal tension members extended over two panel lengths (at times this design is referred to as a Whipple truss, because it was first used by Squire Whipple on the Saratoga and Rensselaer Railroad near Troy, N.Y., in 1852), and the triple-intersection Pratt truss, in which the diagonal members extended over three panel lengths; (2) the Parker truss, in which the top chord is built with a polygonal outline (i.e., the center of the truss is taller than the ends); (3) the camelback truss, a Parker variant in which the polygonal top chord is built with exactly five slopes; (4) the Baltimore truss (named after the Baltimore and Ohio Railroad), in which the panels are subdivided; (5) the Pennsylvania truss (named after the

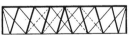

Howe truss, a two-span structure across the Shoshone River near Cody, Wyo., c. 1905.

Post truss, with inclined compression members, near Pine Brook, N.J., c. 1905.

Double-intersection Pratt truss across the Missouri River at Omaha, c. 1910.

Parker truss, a two-span highway bridge near Oscoda, Mich., c. 1940.

Lenticular truss, a two-span deck truss near Plymouth, N.H., c. 1905.

Bowstring arch truss, a high-way bridge in Bristol, Ind., c. 1910.

Pennsylvania Railroad), which is a Parker truss with subdivided panels; and (6) the lenticular truss, in which both the top and bottom chords are polygonal and form a lens shape.

These permutations on the Pratt truss configuration had varied significance, but the names all obtained a modicum of acceptance in the engineering world. Names for patented trusses were commonly applied to structures for which no royalty was ever contemplated or paid. A Pratt truss built in a particular location, for example, in no way implies that Thomas and Caleb Pratt ever received compensation as the result of its construction. In the 19th century patent law and patent enforcement were different from what they are today. In addition, many patents were issued for designs that, strictly speaking, should not have been patentable if U.S. laws had been properly enforced. For example, inventions are not supposed to be patentable if a design has already been built by someone else. However, patent examiners often appeared to ignore this stipulation.

In 1841 Squire Whipple (1804–88) patented the first

Warren truss across the
Skunk River near Colfax,
Iowa, c. 1910.

Warren truss with verticals,
a deck truss highway bridge
near Canby, Ore., c. 1940.

bowstring arch truss. Later, other engineers developed
their own versions of this design, and it was used for
numerous highway spans in the 19th century. Its semicircular shape is similar to a bow, and it consists of a curved
top-chord compression member held together by a
bottom-chord tension member. The vertical tension
members hang from the top chord and help support the
floor beams. Although some engineers might consider
the design more a tied arch than a truss, the latter
designation has achieved widespread acceptance.
Bowstring arch trusses were inexpensive and lightweight, yet sturdy, designs, and for this reason they were
often used for rural highway crossings.

The final major truss type developed in the 19th
century was the Warren truss. Its name is derived from
Capt. James Warren, a British engineer who patented it
(in concert with Theobald Monzani) in 1848. It did not
become widely used in America until the early 20th
century. The Warren truss is distinguished by diagonal
members designed to carry both tensile and compressive
forces. At times, these diagonal members can be supplemented by vertical members to create what are known as
Warren trusses with verticals. Renewed interest in the
Warren truss in the early 20th century coincided with the
development of new technologies used to connect truss
members together.

During most of the late 19th century, American truss bridges used pin connections to hold the various members together. This form of connection required holes to be drilled in the ends of members that were then aligned with one another. Then a cylindrical pin, similar to a large metal dowel, was slipped through the opening to form the structural connection. Pin connections were popular because they allowed for speedy erection of trusses and, theoretically, they made it easier to analyze stresses in a truss. They were also susceptible to loosening, however, especially under the shaking caused by fast-moving, large trains.

In contrast to pins, riveted connections provided a solid, rigid means of joining together the truss members. Riveted connections could not be easily hand driven in

Right: Typical pin connection. Below: Bottom-chord pin connections on a Pratt through truss. The bottom chord is formed by eyebars.

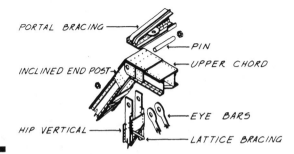

Typical riveted connection, a more rigid means of joining truss members.

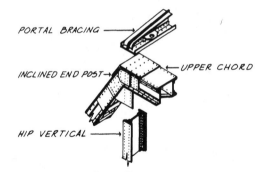

the field, however, and this problem seriously impeded their use for bridge construction until the development of portable pneumatic riveting systems in the late 1880s and 1890s. Warren trusses, which rarely were built with pin connections, became quite popular with the advent of widespread field riveting. Variations using riveted connections also flourished, such as the double-intersection Warren truss and the all-metal lattice truss (similar in design to the wooden Town lattice truss). Beginning in the early 20th century, riveted connections were commonly used for Pratt trusses, although pin-connected Pratt trusses continued to be built in some regions through the 1920s.

In the late 19th century truss bridge technology in America advanced through the efforts of numerous private bridge companies. Concentrated in the northeastern and midwestern United States, these companies often specialized in a few particular types of truss designs and fabricated them for use by cities, counties and railroad companies. Although some of these firms were criticized by engineers as providing unsafe, shoddy products, on the whole they supplied good-quality structures to their clients. By the end of the 19th century, the design of truss bridges was rapidly becoming

Left: Double-intersection Warren truss, a single-span structure in Valley Falls, R.I. Below: Lattice truss, a two-span bridge in Westfield, Mass., c. 1910

Top: Promotional postcard distributed by the Canton Bridge Company, c. 1900. Above: Firth of Forth Bridge (1890) near Edinburgh, Scotland, the world's most famous cantilever bridge. The structure is deepest over the piers.

standardized, and economic pressures precipitated the closing or takeover of many bridge companies. By this time steel had largely supplanted wrought iron as a structural material, and the corporate consolidations accompanying this change further exacerbated the decline of independent bridge companies.

In 1900, for example, Andrew Carnegie set up the American Bridge Company as a subsidiary of the newly formed United States Steel Corporation. Carnegie subsequently bought out more than 25 of the largest bridge companies in America and amalgamated them into the American Bridge Company. Several other bridge companies survived well into the 20th century, but the era of the independent bridge firm was largely over by the end of World War I. In its place, independent consulting bridge engineers provided designs that steel-fabricating firms then erected for whatever company or government agency was paying for the bridge.

CANTILEVER BRIDGES

Most truss bridges in the 19th century were designed so that each span rested independently on its piers or abutments. Known as simple trusses, these bridges did not extend continuously over the piers. For long-span structures (i.e., bridges with long distances between piers), it was economically desirable to design the truss to

run continously over a pier, thus constructing a bridge that would cantilever, or extend, beyond the piers. Like simple truss bridges, cantilever bridges are built using a large number of relatively small members. Visually, they usually differ from their "simple" counterparts in that the trusses get deeper (or taller) where they pass over the piers. This is because they must be designed to resist large bending stresses in these locations. Large cantilevered trusses were occasionally built in the late 19th century, but it was not until after World War I that they were used for numerous long-span highway and railroad crossings.

Top: U.S. Route 40 swing bridge in action across the Susquehanna River, Havre de Grace, Md., c. 1925. Above: Vertical-lift bridge opening to allow passage of a ship through Suisan Bay, Benicia, Calif., c. 1940.

MOVABLE BRIDGES

Rivers and harbors often support large amounts of waterborne commerce, prompting construction of bridges that can be moved to provide navigation clearance. Many movable bridges use designs similar to typical fixed trusses but have special design features to facilitate their movement. They also can be built using simple girders for short spans. Movable bridges can be divided into three basic types: (1) swing, (2) lift and (3) bascule. Swing bridges function just as their name implies: they swing or rotate around a central pier to provide a passageway for ships. Visually, they are distinguished by round central piers and a truss configu-

Rolling lift railroad bridge in the open position across the Cuyahoga River in Cleveland, c. 1910.

ration that appears to be considerably deeper over the pier. Lift bridges provide clearance by having the movable truss span move upward vertically. These tend to be easy to recognize because large towers that house the lifting equipment usually hover above the movable span. Bascule bridges are balanced structures (the name derives from the French term for a weighing device or seesaw) that can be tilted at the abutments to move up and out of the way of ships. Bascule bridges that use mechanisms to roll the structure along its support are known as rolling lift bridges. Others that use hinge mechanisms usually are called heel and trunnion designs. Movable bridges were built throughout the 19th century, but the technology developed into its modern form between the 1880s and the 1920s.

METAL ARCH BRIDGES

Most iron and steel bridges built in the United States in the 19th and early 20th centuries incorporated some type of truss design, but in several instances they were designed as arches. The earliest American metal arch bridge was built on the National Road in southwestern Pennsylvania in the late 1830s. In the 1870s the Eads Bridge in St. Louis demonstrated that long-span arches were less expensive to build and served as a precedent for other structures in the following decades.

Arch bridges were considered aesthetically pleasing and, for certain sites with strong rock foundations, proved economically competitive with other types of designs. Like truss bridges, they can be characterized as through arches or deck arches. Depending on the number of hinges built into the design, arch bridges can be further characterized as fixed, single-hinged, two-hinged or three-hinged. Structurally, the number of hinges is important because it determines how stresses are distributed through the arch. Although the hinges in arch bridges are much larger, they bear a strong similarity to pin connections used for trusses.

Below: First iron bridge in the world (1779), an arch span in Coalbrookdale, England. Bottom: Steel deck arch railroad bridge in Costa Rica, similar to many designs built in the United States, c. 1910.

SUSPENSION BRIDGES

Despite the fact that suspension bridges are built relatively infrequently, some of the longest and most famous bridges in the United States are suspension bridges, such as the Brooklyn Bridge (1883) and the Golden Gate Bridge (1937). As the name implies, the traffic deck of the span is suspended from an iron or steel cable that runs across two tall support towers. To provide for stability this cable is anchored into the abutments at both ends of the crossing using heavy stone or concrete anchorages imbedded in the foundations.

Suspension bridges were first popularized in America in the early 19th century by James Finley. These early designs incorporated wood towers and suspension chains. During the 1840s Charles Ellet (1810–62) and John A. Roebling (1806–69) pioneered the use of iron-wire cables and stone towers. Roebling designed several major suspension bridges, including the world-famous Brooklyn Bridge, and established the standard for the modern type. During the 20th century suspension bridges have been built with steel towers and have produced spans exceeding 4,000 feet. For crossings that

Right: Late 19th-century all-metal suspension bridge with masonry towers, c. 1905. Below: Guyandotte Bridge, Huntington, W. Va., a suspension bridge across the Connecticut River near Brattleboro, Vt., c. 1910.

require extremely wide clearances, they are essentially the only type of bridge that can be built economically. In the early 20th century they were also popular in parts of the South and West for small-scale highway bridges.

REINFORCED-CONCRETE BRIDGES

In the late 19th century bridge engineers began to develop designs using reinforced concrete, a structural material that uses concrete, which is strong in compression, and supplements it with steel, which provides tensile strength. Initially, reinforced-concrete bridges were similar in basic form to stone arch bridges. But gradually engineers began to develop more daring designs that differed in size, scale and form from their masonry counterparts. Although American bridge engineers never proved as adept as European designers such as Robert Maillart in exploiting the structural opportunities offered by reinforced concrete, they nonetheless created some innovative and often beautiful designs.

As with stone, most large reinforced-concrete bridges are arch structures. These can be divided into deck

Below: Open-spandrel, reinforced-concrete arch bridge (1915) in Saskatchewan, Canada, designed by prolific American engineer Daniel Luten. Bottom: Melan Arch Bridge (1897), Topeka, Kans., designed by Edwin Thacher as the first major reinforced-concrete bridge in the United States.

Top: Closed-spandrel, reinforced-concrete arch bridge (1913), Reynoldsville, Pa. The Beaux Arts–influenced balustrade is similar to those used for many early 20th-century spans. Above: Detroit-Superior High Level Bridge (1917), Cleveland, a four-ribbed arch span.

arches, in which the roadway lies on top of the arches, and rainbow (through) arches, in which the arch extends above the roadway like a metal bowstring arch truss. Deck arches were first popularized in America in the 1890s by Fritz von Emperger and Edwin Thacher (1840–1920). Later, Daniel Luten established an extremely successful business building structures of this type through several regionally based construction firms. For spans greater than about 100 feet, open-spandrel arches, which reduced the amount of concrete placed between the arch proper and the road deck, often proved economically attractive. The term "spandrel" refers to the area between the bottom of the arch and the roadway deck of a deck arch bridge. If this area is completely filled in, the bridge is called a closed-spandrel bridge. If it is opened up with only a series of struts or supports connecting the deck and arch, it is known as an open-spandrel bridge. Obviously, using open-spandrel designs

can save considerable amounts of material.

Another means of reducing the concrete in an arch bridge design concerns the form of the arch itself. Instead of building a solid arch that extends the width of the bridge, it is possible to divide the structure into a series of parallel ribs that function as separate arches. Such multirib arches save material because of the open spaces that exist between the ribs.

The rainbow arch bridge is an aesthetically pleasing design popularized by James Marsh (1856–1936) between 1915 and 1930. Because of his active promotion of the design, it is often referred to as a Marsh arch, especially in the Midwest. Other engineers, however, also built similar designs. As with open-spandrel designs, the rainbow arch bridge was economically attractive because it conserved materials.

Reinforced-concrete bridges became popular because, at least initially, they reduced maintenance costs. In addition, they made use of locally available materials such as sand, gravel and, in many areas, cement, and they could be built by relatively unskilled laborers. Thus, by building a concrete bridge a city or county engineer could keep construction funds in the local economy rather than send them to a steel company in another state or region.

Top: Massive example of a reinforced-concrete rainbow arch design, Ridgway, Pa., c. 1925. Above: Promotional card distributed by the Marsh Engineering Company, c. 1920, showing an example of the company's work across Wakarusa Creek, Shawnee County, Kans.

In addition, many people consider reinforced-concrete arch bridges visually more attractive than comparable steel trusses. For this reason they were often selected for sites in picturesque locations or in fashionable urban areas.

GIRDER BRIDGES

One of the most widely used yet least appreciated types of bridges is the girder. In simple terms, girders are solid beams that extend across a small-span crossing. In the 17th and 18th centuries wood was used, while later girder bridges called for iron, steel and reinforced concrete. By the early 20th century they were often used for railroad structures because they could provide solid, stable crossings capable of withstanding fast-moving, heavy traffic. Their major liability was that they required extensive amounts of material, especially compared to truss designs of similar size. For short structures, however, this cost difference was not particularly significant. Steel girder bridges are usually formed by riveting together large steel plates; hence, they are often called plate girders.

With the growth of highway systems after World War II, steel and concrete girder bridges became the most common type of bridge built in the United States. Although often overlooked by bridge historians, the girder bridge has been around for hundreds of years and

will continue to be the most common type of bridge in America for decades to come.

Bridges of all types and all eras tell us much about America's past — its transportation and economic needs as well as the engineering challenges that have been overcome in the growth of our republic. The site descriptions in this book are intended to increase understanding of bridge technology and foster an appreciation of how bridges contribute to America's social and economic development.

Above: Steel-plate girder highway bridge near Rockford, Ill., c. 1905. Left: Reinforced-concrete girder approach spans, built as extensions of the San Francisco–Oakland Bay Bridge, San Francisco, 1935.

DAMS: CONTROLLING A PRECIOUS RESOURCE

Water is a physical necessity for human survival, and, in arid environments, collection and storage of the "precious liquid" are vital prerequisites for cultural development. The oldest known dam in the world is a small earthen and rockfill structure in the Jordanian desert that dates to the third millenium B.C. Dams were built by many cultures in the Middle East and played a key role in the growth of the irrigation-based societies of Mesopotamia. The Romans were prolific dam builders in the Mediterranean region (some of their structures are still in active use), and Spain subsequently became an important focus for dam construction in the Middle Ages and the Renaissance. By the 18th century Spanish engineers were beginning to rationalize the process of designing and building dams, and they prepared some written treatises on the subject. This codification of dam-building formulas was further developed by French and British engineers in the late 18th and early 19th centuries. Although intuition and experience continued to play an important role in the art of dam construction, by the late 19th century mathematical analysis began to constitute a critical aspect of the design process in both Europe and the Americas.

DIVERSION AND STORAGE DAMS

In functional terms dams can be divided into two basic categories: diversion and storage. Diversion dams are usually relatively small structures built to divert or deflect water from a stream or river into a specially built canal or conduit. Structures of this type can be designed to divert the entire flow of water in a waterway or, as is most common, only a portion of the flow. As a result, diversion dams are usually constructed so that they can be overtopped during times of heavy flooding. "Overtopping" means that the height of water in a reservoir is taller than the dam and, consequently, water flows over the top of the structure.

While the main purpose of a diversion dam is to redirect the flow of water for use elsewhere, storage dams are designed to retain water for long-term use. They are built not to divert water (although this can be a secondary function) but to impound seasonal runoff for use throughout the rest of the year. In many regions most of the yearly precipitation comes in the spring and early summer in the form of rain and melted snow, which swell river levels for a few months. During the late summer and the fall, the flow in these rivers can drop off to a mere trickle. Building large storage dams makes it possible to impound spring floods and, during the latter part of the year, distribute this water to downstream users. Storage dams are usually much larger than diversion dams and, therefore, more expensive to build.

Opposite: Hoover Dam (1935) across the Colorado River on the Nevada-Arizona border. The dam provides hydroelectric power for the Southwest and stores water for domestic and agricultural use in Arizona and southern California.

Above: Simple wooden diversion dam across the Susquehanna River near Windsor, N.Y., c. 1910. Right: Reinforced-concrete curved gravity storage dam (1919) near Tapoco, N.C., built for hydroelectric power generation.

The initial use of a region's water resources typically involved the construction of relatively small-scale diversion dams. As time passed, making use of floodwaters became more economically desirable; consequently, the construction of storage dams would then begin to figure more prominently in regional development. In addition, it is worth noting that, although storage dams are usually larger than diversion dams, the various dam types listed below can generally be adapted for either function.

PUTTING DAMS TO USE

Dams serve many purposes. The oldest of these uses are domestic water supply and irrigation. The first involves the control of water to serve an elemental need of human existence; the second provides for the artifical watering of agricultural crops in arid environments. The early hydraulic civilizations in Mesopotamia and other parts of the Middle East were all based on the development of irrigated agriculture, and their water control technologies included the use of dams. As noted in several site descriptions in this guide, modern irrigation in the American West continues to depend on the use of storage

Left: Wachusett Dam (1905), Clinton, Mass., a storage dam built as part of Boston's municipal water supply system. Domestic water supply is a critical function of many storage dams. Below: California orange grove being irrigated, c. 1910.

and diversion dams. In general, irrigation involves diverting water out of a river and having it flow into a man-made canal. This canal declines at a slower rate than the river itself, so that after a few miles the canal is able to pass through land that is well above the flood plain of the river. By releasing water onto the land via laterals (branch canals), it is possible to nourish crops on otherwise dry land.

Beyond meeting domestic and agricultural needs, the most important service provided by dams is power production. The development of water power — the conversion of the kinetic energy within falling water into mechanical power— is evident in its most simple form in small-scale grist mills used to produce flour and other products essential in an agricultural economy. By the mid-19th century overshot and undershot waterwheels began to be replaced by smaller and more efficient turbines. After the development of electric generators in the late 19th century, water-powered turbines were quickly adapted to generating hydroelectricity. Turbines are able

Long Lake Dam (1915) in Washington State showing the hydroelectric power-house on the right. Water from the reservoir flows through pressurized pen-stocks to reach turbine-generator units in the powerhouse.

to convert efficiently the kinetic energy of falling water into rotary motion; electricity is then created by attaching the spinning or rotating turbines to electric generators. An electric current is produced when a conductor (usually made of copper) is moved through a magnetic field. The spinning of a generator powers the constant movement of a conductor through such fields, and this is the source of electricity. Dams can facilitate hydroelectric development by increasing the head (height) of water that acts on the turbines, and this in turn increases the power potential of the system. Dams can also store floodwaters so that power production can be maintained year-round, even during normally dry periods.

Dams have been used to store water for myriad other uses, such as pressurized hydraulic mining, the creation of artificial logging ponds and water supply for transportation canals. Most of the dams described in this guide, however, are used for domestic water supply, irrigation or hydroelectric power generation. At least one was developed exclusively to provide flood control, and a few were built to raise water levels along major rivers and, thus, increase the economic feasibility of inland navigation.

GRAVITY DAMS

Dam technology can be separated into two general traditions: the massive and the structural. The first is the most prevalent and also the simplest to comprehend. Massive dams resist hydrostatic water pressure (the pressure exerted by a volume of water) by the sheer mass (bulk) of the materials. The underlying principle of massive dam construction is to build up sufficient quantities of earth, rockfill, masonry or concrete so that the pressure of the stored water is insufficient to push the dam downstream. In essence, the force of gravity acting on the dam is what provides structural stability. Consequently, dams of this type are commonly called gravity dams.

The earliest gravity dams consisted of earth, rock or timber or a combination of these elements. Their design

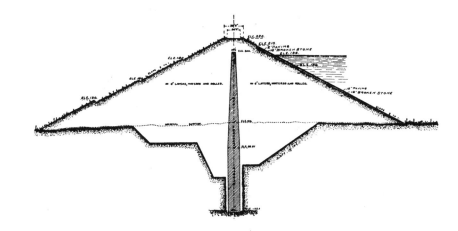

Above: Triangular cross section of an earthfill design with a masonry core wall proposed by the New York Aqueduct Commission, c. 1900. Left: Grassy downstream slope of the earthfill Gainer Memorial Dam (1927) in Rhode Island. Below: Timber crib dam near Dover, Maine, c. 1910.

was empirically based, and if insufficient material was placed in a structure, it would wash away. Although structures of this type were quite susceptible to being washed out, especially during times of heavy flooding, they remained popular because they often used readily available local materials. The shape of massive dams was dictated primarily by the structural properties of the material used in their design. It is essentially impossible, for example, to build a stable earthen wall with a vertical face; the earth will naturally collapse. As a result, earthfill dams are necessarily built with sloping sides that form a triangular cross section. Rockfill designs can have slightly steeper slopes, but, like their earthfill counterparts, they must use a wide-base, triangular cross section to maintain stability.

Although loose rockfill cannot be constructed with a vertical upstream face, stone structures carefully built by skilled masons can be erected in such a form. By the early 19th century European engineers were beginning to develop mathematical formulas for determining how much stone masonry was required in a gravity dam to impound safely a given height of water. In the 1850s M. De Sazilly, a Frenchman, postulated an ideal cross-sectional profile of a vertical-faced masonry gravity dam based on a typical density of stone (approximately 150 pounds per cubic foot) and water that weighed 62.5 pounds per cubic foot. By adding together the horizontal vector, representing the hydrostatic pressure, and the vertical vector, representing the force of gravity acting on

Below: Escondido rockfill dam (1895), north of San Diego. Bottom: Hiwassee Dam (1940) in western North Carolina, a typical straight-crested concrete gravity dam.

the mass of masonry, it became possible to calculate a resultant force acting on the base of the dam. As long as this resultant force fell within the middle third of the dam's base, then the structure would maintain stability and not tip over. De Sazilly's rational profile, also called the profile of equal resistance, was basically triangular in shape with a height-to-width ratio of 3 to 2.

Other engineers refined De Sazilly's rational profile for gravity dams, but its basic form remained popular for many years, continuing even today. Used for large masonry designs in the late 19th century, it was adapted for concrete gravity structures in the 20th century. Although masonry and concrete gravity dams use large amounts of material (and, therefore, can be quite expensive to build), they are relatively simple to design and usually do not present complicated construction problems.

Masonry gravity dams built straight across a valley or canyon are known as straight-crested gravity dams, while those built along an upstream curve are known as curved

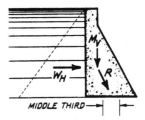

WH = Horizontal Water Pressure
WY = Vertical Water Pressure
MV = Vertical Dam Mass
R = Resultant Vector

Cross-sectional diagram showing forces acting on a vertical-face gravity dam.

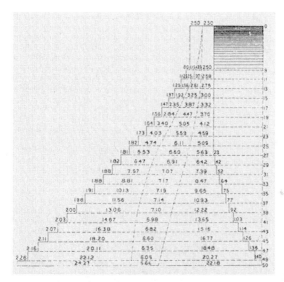

Left: "Rational" gravity dam profile developed by De Sazilly in the 1850s. Below: Roosevelt Dam (1911) near Phoenix, a masonry curved gravity dam.

gravity dams. Curving a gravity dam has been thought to make it stronger, but this is not necessarily true. The dam continues to function primarily as a gravity structure and not as an arch. Curving the dam certainly makes it appear stronger, and this psychological factor has played a role in the selection of curved dams for many large projects. Some engineers also preferred to curve gravity designs because they believed this reduced the tendency of shrinkage cracks to appear in the concrete.

ARCH AND BUTTRESS DAMS

In contrast to massive-dam technology, the structural tradition of dam design does not rely on the sheer bulk of material in a dam to provide stability. A structural dam uses substantially less material than a comparable gravity dam by depending on its shape, not just its size, to hold back water. Structural dams fall into two basic types: arch dams and buttress dams.

Arch dams may appear similar to curved gravity dams, but they are much thinner than their massive counterparts. An arch dam has a cross-sectional profile that would not be stable if used for a gravity design. The curve of its upstream face is a critical aspect of arch dam design: the arch transfers the hydrostatic pressure acting on the upstream face of the dam to the canyon walls on both sides of the dam site. Arch dams are not suitable for all dam sites because they require relatively narrow canyons with solid bedrock foundations. Arch designs frequently use much less material than comparable gravity designs, however, and for many sites they can be considerably less expensive to build.

Arch dams were built as early as Roman times and, later, in Persia and Spain. But it was not until the mid-19th century that the French engineer Emile Zola developed a simple mathematical formula for analyzing basic stresses in arch dams. The first major American arch dam was a 64-foot-high design built in Bear Valley in southern California during the 1880s. This structure is no longer in use, but it inspired other engineers to innovate with designs that depended on more than mere bulk to provide stability. Some early arch dams were constructed of masonry, but by the early 20th century concrete came into almost universal use.

Like arch dams, buttress dams also require much less material than comparable gravity structures, but they achieve their material efficiency in a different way. Rather than employ a vertical upstream face, buttress dams usually have an upstream face that slopes into the

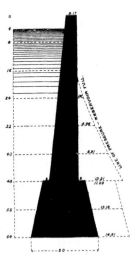

Cross section of the Bear Valley Dam (1884) in southern California. The arch structure is much too thin to function as a gravity dam.

Cross-sectional diagram showing forces acting on the inclined upstream face of a buttress dam.

WH = Horizontal Water Pressure
WY = Vertical Water Pressure
MV = Vertical Dam Mass
R = Resultant Vector

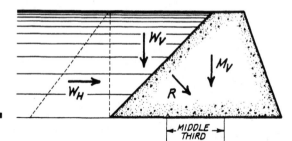

Mill Dam (c. 1900) across the Blackstone River, Valley Falls, R.I. Although it appears to be an arch dam, the dam is actually a curved gravity structure. The curved design increases the dam's crest length and reduces the depth of overflow floods.

F. E. Brown's Bear Valley Dam (1884), a true arch dam of daring dimensions. The hard rock foundation at the site is required for arch dams.

Construction of a reinforced-concrete, flat-slab Ambursen buttress dam on the Juniata River near Huntingdon, Pa., in 1906. Note the hollow nature of the overflow design.

reservoir. This means that the water in the reservoir exerts both a horizontal and a vertical load on the dam. As with a gravity design, the stability of the structure requires that the combined force of the water pressure and the weight of the dam proper pass through the center third of the base. But, because the vertical component of water pressure acts on the upstream face of the dam, not as much material is required to build a stable structure.

In essence, buttress dams act as gravity structures but are designed to take advantage of the vertical component of water to achieve structural stability. The amount of material in the dam is reduced by building a series of discrete buttresses spaced from 15 to 70 feet apart. The inclined upstream face is then built across the front of these buttresses. When the reservoir fills up, the water pressure acting on the upstream face is concentrated on the buttresses, which carry this pressure down to the foundation. The downstream side of a buttress dam does not appear as a solid monolith. Instead, it usually appears as a series of parallel walls. As a result, buttress dams are often referred to as hollow dams because of the hollow spaces between the buttresses. Although they have been built of wood, steel and masonry, the vast majority of buttress dams use reinforced concrete.

Depending on the shape of the upstream face, buttress dams generally can be placed in two categories: flat-slab dams and multiple-arch dams. As the name implies, the flat-slab design has a flat upstream face that connects the buttresses. In a multiple-arch dam the upstream side is formed by a series of arches supported on the buttresses. During the early 20th century both flat-slab and multiple-arch buttress dams were built using reinforced concrete, and in many ways the two technologies are

quite similar. Multiple-arch dams, however, allowed for wider spacing between the buttresses, so they used less material than a flat-slab design of comparable size. The earliest flat-slab buttress dams were designed by Nils F. Ambursen and his company, the Ambursen Hydraulic Construction Company; hence, they are often called Ambursen dams. The first major multiple-arch dams were built by John S. Eastwood (1857–1924), a California engineer; the technology was subsequently adapted by several other engineers in the United States and Europe.

Upstream face of the Big Bear Valley reinforced-concrete, multiple-arch buttress dam (1911) in southern California. Built by John S. Eastwood (inset), this structure replaced F. E. Brown's 1884 arch dam visible in the foreground.

DAMS IN THE UNITED STATES

In America the first dams of the 18th and early 19th centuries were relatively small, crude timber crib and stone diversion dams used to power grist and saw mills. In this context, "crib" refers to a three-dimensional matrix of timbers that could be filled with stones, rocks and dirt. With the development of large water-power sites and urban water supply systems in the mid-19th century, dams became the object of more focused engineering design. During this period, the tradition of massive dam design held sway throughout the eastern United States. The opening up of the arid West to settlement in the late 19th century provided a new impetus for dam construction, as the need to store water for mining, irrigation and hydroelectric projects in isolated, remote western locations fostered interest in more innovative and less expensive designs. Although massive gravity designs were frequently used in the West, the region became a stronghold of the structural tradition during the early 20th century. In response to the Depression of the 1930s, Franklin Roosevelt's New Deal used dam construction as

William Gropper's 1937 study for a mural celebrating dam construction projects during President Roosevelt's New Deal.

a major means of reviving the West's faltering economy, and numerous concrete and earthfill gravity dams were built, including the Bonneville Dam in Oregon, the Grand Coulee Dam in Washington and the Fort Peck Dam in Montana.

With few exceptions, since World War II practically all dams built in the United States have been expensive massive gravity structures, such as the Oahe Dam in South Dakota. The reasons for this are complex and relate to numerous factors, including the nature of the federal appropriations process and the development of large-scale, earth-moving and concrete-conveyance technologies. While the structural tradition of dam design has been largely dormant in America for several decades, it was responsible for many important dams in the early 20th century — such as the Salmon Creek Dam in Alaska and the Littlerock Dam in California — and made significant contributions to the development of the nation's water resources.

The history of dams is extensive, of course, but one of the best ways to learn about dam design and the role of water impoundment structures in American history is to look at a variety of sites throughout the United States. The entries that follow in the guide section are a good place to start.

Construction of the Chief Joseph Dam near Bridgeport, Wash., in 1954. This federally funded structure is typical of many concrete gravity dams built after World War II.

Downstream face of the Army Corps of Engineers' Bonneville Dam (1937), a concrete gravity overflow structure on the Columbia River, east of Portland, Ore.

SAVING BRIDGES AND DAMS

As engineering structures, bridges and dams face much greater preservation problems than buildings. Bridges and dams are built to serve the utilitarian needs of transportation and water supply. Although they may possess great aesthetic power and are often important parts of a community's cultural heritage, their preservation as historic structures is dependent on contemporary engineering issues that usually bear little relation to history. The preservation and continued use of historic bridges and dams can be achieved only if (1) engineering authorities responsible for the structures can be convinced that they are safe and do not pose any danger of collapse or destruction; and (2) they are adequate to meet modern needs or are located in a place where their preservation would not impede the growth of a larger transportation or water supply system.

Of course, public safety is an important factor in the preservation of all old structures, and many historic properties have been demolished because of concerns over structural stability. With bridges and dams, however, safety is of such great importance to engineers, government officials and the public that preserving structures because of their historical significance often is given scant thought.

Group of boys finding an afternoon of amusement at the Doan Brook Dam, Cleveland, c. 1905.

■ ■ ■ ■   BRIDGE PRESERVATION   ■ ■ ■ ■

In the early 1970s, public interest in the value and significance of America's historic bridges began growing. With the lead of enthusiasts interested in wooden covered bridges, historic bridges came to be seen as artifacts representing important developments in structural technology and as vital components of historic districts, towns and regions. At the same time, plans to upgrade America's highway system were expanding, and preservationists became aware of the enormous number of historic bridges scheduled for demolition with both state and federal funds. Highway officials have urged the replacement of thousands of the nation's 575,000 bridges during the next decade and beyond. Some states have been replacing several hundred spans each year, and one state engineer has expressed a desire to replace 500 annually. In 1987 the Federal Highway Administration told Congress it needs more than $50 billion to replace and repair about 220,000 bridges.

As with other aspects of historic preservation, the circumstances surrounding the preservation of specific bridges are often unique. Certain issues and principles, however, are relevant to a wide range of bridge preservation situations. The most important element in the successful, long-term preservation of historic bridges is strong local support; and, because most of the laws relating to bridge preservation concern activities funded

Opposite: San Francisco's Golden Gate Bridge during the celebration of its 50th anniversary in 1987. More than 200,000 people jammed the roadway.

Restoration of one of the Kennedy family's 19th-century covered bridges in Rush County, Ind. Rush County Heritage has helped save six of these landmarks.

or authorized by the federal government, this support must be of sufficient visibility to impress regional, state and federal officials. An understanding of the importance of historic bridges, combined with active public support for using federal programs to encourage preservation, can be the key to increasing rehabilitation and reuse of historic bridges.

REPLACEMENT VERSUS REHABILITATION

Federal support for the development of America's highway system first appeared in the early 20th century. Research into the durability of various highway components and the development of standards for state and local engineers to follow in highway construction projects were early concerns of the Bureau of Public Roads (now the Federal Highway Administration or FHWA). With World War II and the expansion of America's car culture, highway growth became inextricably associated with both defense needs and the general health of the national economy. Since the 1950s and the initial construction of the Interstate Highway System, the Federal Highway Administration has administered all aspects of the federal-aid highway system in conjunction with state highway departments.

The problem of bridge replacement did not receive special attention until the 1967 collapse of the Point Pleasant Bridge (1928) over the Ohio River, in which 46 people died. The publicity generated by this bridge failure is often credited with justifying the inclusion of a Special Bridge Replacement Program in the Federal-Aid

Highway Act of 1970, which provided for demolition of substandard bridges and the subsequent construction of new ones. It authorized no funds for the maintenance, rehabilitation or restoration of existing structures.

In 1978 Congress added funding for rehabilitation through the creation of the Highway Bridge Replacement and Rehabilitation Program. From 1978 through 1986 this program made $11.1 billion available to replace or rehabilitate bridges both on and off the federal-aid highway system; another $8.15 billion was authorized for work through 1991. In the regulations implementing the bridge program, "rehabilitation" is defined as "the major work required to restore the structural integrity of a bridge as well as work necessary to correct major safety

Left: Large multispan bridge near Florence, Ala., built in the 1930s as part of America's expanding public highway system. Below: Repair work on the Brooklyn Bridge (1883) in the 1930s.

defects." In the past, highway engineers have almost always interpreted this to mean that rehabilitation of a bridge requires it to be brought in line with standards developed by the American Association of State Highway and Transportation Officials. AASHTO standards usually require that a new bridge's roadway deck exceed 30 feet in width, that it be capable of carrying heavy truck traffic and that its approaches and alignment be as close to a straight line as possible. It can be difficult for a historic bridge to be rehabilitated in accordance with such standards, and almost all projects funded by the Federal Highway Administration's bridge program have totally replaced old spans with new structures.

With the enactment of the Highway Improvement Act of 1982, the threat to surviving historic bridges became even more serious. This act, which reauthorized all the federal highway programs, provided the bridge program with billions of dollars. In 1987 the program received even more funding under provisions of the Surface Transportation and Uniform Relocation Assistance Act, passed by Congress over a presidential veto. Along with increased funding for bridge replacement programs, however, the 1987 act also provided a major boost to efforts to preserve historic bridges. With key support from Sen. Robert T. Stafford of Vermont, chairman of the Senate Committee on the Environment and Public Works, the act established it "to be in the public interest to encourage the rehabilitation, reuse and preservation of bridges significant in American history, architecture and culture. Historic bridges are important links to our past, serve as safe and vital transportation routes in the present, and can represent significant resources in the future." Beyond statements of policy, the new law (1) required all state highway departments to undertake an inventory of historic bridges (if they had not done so already); (2) clarified that federal funds could be used to rehabilitate historic bridges for vehicular use or for such purposes as bicycle routes or pedestrian ways; and (3)

Sunday school group posing for a reflective portrait on a Pratt pony truss near Leipsic, Ohio, c. 1910. Bridges can serve unanticipated roles in a community's culture.

allowed money that would be necessary to demolish a historic bridge to be made available to state, local or private entities so they could either move the structure or take responsibility for maintaining it in its present location. As the National Trust for Historic Preservation pointed out in seeking enactment of these changes, this latter provision "will not cost the federal government any additional money, and it may enable governments or preservation organizations to own and protect historic bridges." Despite the increased federal funding available for replacing old bridges, the new law holds out hope that at least some historic bridges will find new uses in the years ahead.

Regulations implementing the bridge program provide that states may participate by conducting bridge inspections and submitting inventories to the U.S. Department of Transportation through the Federal Highway Administration. A numerical sufficiency rating from zero to 100 is then assigned to each bridge in accordance with a formula developed by the FHWA and sanctioned by AASHTO. Any bridge rated less than 50 is eligible for the bridge program.

Despite flexibility allowed by FHWA regulations, the practical effect of strictly applying AASHTO standards has been to encourage the demolition of historic structures, although this is not the necessary result of the requirements of the federal bridge program. FHWA regulations specifically allow for exceptions, on a project-by-project basis, for designs that do not conform to the minimum AASHTO design criteria. Appeals for exceptions can be made to the FHWA state administrator through the state department of transportation.

Highway bridges can be determined unsafe for two general reasons. The first relates to what highway engineers term "functional obsolescence." This means that the bridge is no longer considered adequate to serve modern transportation needs. A bridge can be functionally obsolete because it is too narrow to meet modern standards (a frequent problem with historic spans) or because the alignment of the structure results in an unsafe curve in the highway right-of-way. A bridge can be in perfect structural condition but be considered unsafe because of either width or alignment problems. Narrow, poorly aligned structures can have good safety records because local residents are aware of their limitations and slow down when approaching them. However, highway engineers are usually unimpressed with statistics of this type when they are planning highway improvements.

The second type of safety problem encountered with old bridges concerns structural deficiencies. These occur when some part of the bridge has deteriorated because of rust, cracking or decay. Analysis also may demonstrate the original design to be inadequate for modern loads. Given how much heavier trucks have become since World War II, it does not take much imagination to conclude that most bridges more than 50 years old that have received only modest maintenance can be classified as structurally deficient.

In terms of catastrophic failure, most bridges collapse because of overloading by a truck that is heavier than the legal posted limit or because a vehicle runs off the roadway and hits a key component of the structure. At times, bridges can fail suddenly because of deterioration (usually rust) that weakens a critical part of the structure. This was the cause of the failures of the Point Pleasant Bridge in 1967 and the I-95 span over the Mianus River near Greenwich, Conn., in 1983. Bridges can also fail if their foundations are eroded by flooding. This was a frequent cause of railroad bridge failures in the late 19th century, and it caused the tragic collapse of the New York State Thruway bridge over Schoharie Creek in 1987.

Pratt through truss in Ottawa, Kans., weighted down with rocks to help stabilize it during a 1915 flood. Floods have claimed many bridges through the ages.

PROTECTIVE LEGISLATION

Legislation adopted by Congress directs federal agencies to consider the impact that their programs and activities have on historic and cultural resources. In the case of the federal bridge program, three pieces of legislation apply:

1. Section 106 of the National Historic Preservation Act of 1966, as amended, requires federal agencies to take into account the effect of their proposed undertakings on properties listed in or eligible for inclusion in the National Register of Historic Places before the expenditure of federal funds or the issuance of any licenses. The federal agency must also allow comment from the Advisory Council on Historic Preservation, a 19-member body established by the National Historic Preservation Act to advise the president and Congress on matters relating to historic preservation. In addition, agencies whose actions may directly and adversely affect a National Historic Landmark must make every effort to minimize this harm.

2. Section 4(f) of the U.S. Department of Transportation Act of 1966, as amended, provides that the secretary

of transportation shall not approve any program or project that requires the use of any land from a historic site of national, state or local significance unless there is no feasible and prudent alternative to the use of such land and all possible efforts are made to minimize harm to the historic site.

3. The National Environmental Policy Act of 1969 is best known for its requirement of an environmental impact statement (EIS) on proposed major federal actions that will significantly affect the quality of the environment, including important historic, cultural and natural aspects of our national heritage.

In spite of their apparent strength, the National Historic Preservation Act, section 4(f) of the Department of Transportation Act and the National Environmental Policy Act have often proved ineffective in halting the demolition of historic bridges. In many cases this weakness may be jointly attributed to so-called paper compliance by the federal agency and the failure of preservationists to use the laws effectively. The procedural mechanisms provide for the participation and involvement of individuals, organizations and governmental entities in the federal planning and decision-making process. Consequently, effective use of these procedural tools requires that preservationists get involved early to help shape the ultimate decision. And, there is encouraging news for preservationists in the 1986 amendments to highway legislation discussed earlier. These changes, by encouraging flexible use of federal bridge funds to preserve historic spans, should give government officials a greater number of alternatives to demolition — alternatives they must consider as required by the protective laws discussed above.

PRESERVATION CHECKLIST

To prevent the loss of historic bridges, the following efforts can be crucial in making the best use of the legal mechanisms mentioned previously:

■ Identify important bridges that may be threatened early on and have them listed or declared eligible for listing in the National Register as well as in appropriate state and local registers.

■ Build strong local support and coalitions and make sure such support is visible to public decision makers through lobbying efforts and the news media.

■ Examine alternatives and develop a preservation plan. An effective way to accomplish this is an engineering feasibility study. It is critical that a strong economic and technical case be made for saving a historic bridge, either in combination with a new structure or on its own. It is highly desirable that such studies document the financial savings that can accompany the reuse of a historic span. This would be the appropriate context in which to consider and propose flexible application of the AASHTO standards.

■ Work closely, even before demolition is proposed, with

local government staff and officials, the state highway department, the state historic preservation officer and any other relevant state or federal agencies to gain support for preservation plans.

■ Participate actively and effectively in the federal preservation review processes, making sure that all of the information gathered, alternatives developed and arguments structured are made a part of the section 106 process, the section 4(f) evaluation and statement, and the environmental impact statement. These data must be incorporated into the EIS and 4(f) statement when they are written, not after the documents become final.

Early identification, planning and building of local support and coalitions have led to a number of successful bridge reuse and restoration projects throughout the country. The following case studies show that creativity, compromise and commitment can help prevent the loss of important examples and offer positive solutions for the protection of historic bridges.

PLANNING BASED ON INVENTORIES

Until recently, one of the most difficult obstacles to bridge preservation efforts was a lack of knowledge about the actual number and nature of historic bridges in a state or region. Bridge inventories undertaken to assess bridges' historical significance (as opposed to their structural adequacy) can yield useful information in determining which bridges warrant preservation interest. In 1982 state highway departments were authorized to conduct historic bridge inventories using federal funds if they so desired; in 1986 the new highway law mandated them. At present, almost all states have completed or are in the process of completing historic bridge inventories.

Most of these surveys have been undertaken as cooperative efforts between state highway agencies and state historic preservation offices. In addition, many states have established multiagency, multidisciplinary advisory committees to oversee the process. For example, a council on historic bridges formed in West Virginia includes members of the departments of highways, culture and history (SHPO) and natural resources as well as the inventory consultant and members of a citizens advisory committee. The goals of this cooperative body included developing an evaluation procedure and priority list of historic bridges that are in need of restoration and preservation; nominating bridges to the National Register; reviewing and approving maintenance and restoration plans for all bridges designated for preservation; and providing professional services in the development and implementation of bridge preservation work. Whether historic bridge inventories are carried out on a countywide or statewide basis, by preservationists or highway officials, they can be useful planning tools. However, the identification of historic resources is only the first step in the bridge preservation process, and it must be followed by an analysis focusing both on which bridges are desirable to preserve and which bridges are feasible to preserve. Unfortunately, these two categories do not always coincide neatly, but an inventory establishes a context in which to make rational preservation decisions. Bridge surveys have been undertaken using a wide variety of methodologies and goals. The following examples illustrate two early surveys conducted at the state and county levels.

The first statewide truss bridge inventory project was undertaken by the Virginia Highway and Transportation Research Council in the early 1970s. More than 500 metal truss bridges were inventoried and the results disseminated in a series of illustrated publications. Because the VHTRC is associated officially with both the University of Virginia and the State Highway and Transportation Commission, its study of metal truss bridges was respected by highway officials. This respect was translated into action when the commission officially supported the nomination of seven truss bridges to the National Register of Historic Places and acknowledged that another 20 to 30 more were of considerable historical significance and warranted special consideration in terms of replacement or rehabilitation.

Virginia's historic bridge inventory played a major role in focusing attention on the large number of historic bridges that face demolition and on the difficulties that state highway departments can have in reconciling historical interests with safety needs. The inspiration of highway engineer Howard Newlon, the inventory also demonstrated that state highway departments could work to help preserve historically significant bridges.

■ **Virginia Metal Truss Bridge Inventory**

Pratt through truss (1882) in Nokesville, Va. This bridge was identified in the Virginia metal truss bridge inventory and subsequently was nominated to the National Register of Historic Places.

Double-intersection Pratt
through truss (1878) on
Poffenberger Road, the
oldest surviving truss bridge
in Frederick County, Md.
Thanks to the efforts of historian Cherilyn Widell, it is
listed in the National Register of Historic Places.

■ **Frederick County, Md.,
Metal Truss Bridge
Inventory**

The value of a local or regional historic bridge inventory is demonstrated by efforts in Frederick County, Md. Located about 50 miles northwest of Washington, D.C., and encompassing more than 600 square miles, Frederick County is a productive agricultural center that is rapidly becoming a residential satellite of metropolitan areas to the east. In 1977 the local county historic preservation commission (with financial support from the Maryland Historical Trust) initiated an inventory of metal truss bridges throughout the county. This inventory identified more than 40 metal truss bridges dating from the 1870s through the 1930s. The inventory coordinator then met with the county engineer to discuss future highway bridge construction plans in the region. Because of the county's interest in retaining its rural character, the county engineer and the county historic preservation office were able to agree on a plan allowing for the preservation and maintenance of many historic bridges under the county's jurisdiction. At the same time, it was agreed that several historically significant bridges would need to be replaced because of their structural condition or because planned highway construction projects would render them obsolete. In essence, a compromise was reached allowing the county engineer to maintain a safe and efficient highway system while also providing for the long-term preservation of many historic bridges on roads that are not critical components of the region's highway system.

USING FEDERAL REHABILITATION FUNDS

Historic bridge inventories are not always necessary for individual historic bridges to be included in a community's long-term transportation system, as the next two examples show.

The Elm Street Bridge in Woodstock, Vt., a Parker truss, became the focus of a confrontation between preservationists and highway interests in 1975–76, and the compromise decision related to its replacement served to promote the feasibility of historic bridge preservation on a wider scale. Located within a National Register historic district, the span highlighted the importance that communities can assign to historic bridges and demonstrated the impact that organized citizens can have on the highway planning process.

The controversy surrounding the bridge occurred because its width and alignment did not meet modern highway design standards. As the bridge was listed in the National Register and the state planned to use federal money for its demolition or alteration, compliance with section 106 of the National Historic Preservation Act, the National Environmental Policy Act and section 4(f) of the Department of Transportation Act was necessary. Through the processes set up by these laws, alternatives allowing retention of the bridge were considered. A memorandum of agreement between the Advisory Council on Historic Preservation, Federal Highway Administration and state historic preservation office resulted in a precedent-setting compromise solution. The bridge and original stone abutments were widened enough to satisfy highway officials, although not enough to meet American Association of State Highway and Transportation Officials standards. Four-foot steel girders were inserted beneath the deck to increase its load-bearing capacity, and the original trusses were mounted on the new deck of the bridge. New details were designed carefully to complement the old. Although the design solution was expensive, the total cost was less than that estimated for completely new construction.

As mentioned previously, the current standards for highways as outlined in the federal regulations do allow for deviations from the AASHTO standards, where unusual conditions warrant. However, the Elm Street Bridge was the first case in which federal highway funds were used on a project that did not meet AASHTO standards, and the waiver was granted because of

■ **Elm Street Bridge**
Woodstock, Vt.
National Bridge and
Iron Company
1870, 1979

Above: University of Vermont preservation students measuring the Elm Street Bridge in the mid-1970s to document its significance. Below: Woodstock's bridge before it was widened and realigned.

historical considerations. According to a former Vermont state historic preservation officer, William Pinney, this occurred because preservation proponents were able to make a strong argument for the functional importance of a smaller bridge that would sufficiently slow highway traffic into town and on the narrow village streets. Studies of the safety record of the 18-foot-wide bridge showed that it was unusually safe—no fatalities in more than 100 years and few accidents. Strong, well-organized local support, as well as the comments from the Advisory Council on Historic Preservation, convinced state and federal highway officials that this was a case where standards could be waived without endangering the motoring public. From a purist point of view the historical integrity of the bridge suffered when it was upgraded, but these alterations did not destroy its visual qualities, especially as seen from the roadway.

■ **Second Street Bridge**
Allegan, Mich.
King Iron Bridge and
Manufacturing Company
1886

The use of federal preservation legislation is not always necessary to save a historic bridge. At times successful pressure can be brought to bear in other ways, as was the case with the Second Street Bridge in Allegan, Mich., a single-span, double-intersection Pratt through truss. Located near the town's central business district, the Second Street Bridge is not a component in any major transportation corridor. Instead, it serves only as a release valve for local traffic leaving the downtown. Because the bridge carries a relatively small amount of traffic, in the late 1970s city officials did not see the need to totally replace the 1886 span with a new structure. To them it made more sense to restore the 18-foot-wide bridge for one-way traffic in a manner that ensured its safety while at the same time using as many of the

Second Street Bridge being
moved back to its site after
rehabilitation in 1983.

bridge's original members as possible. Such a restoration would be cheaper than a new bridge and would meet the public's needs, but at the same time it would not meet AASHTO standards and, consequently, it would not qualify for federal funds.

Convinced that restoration rather than replacement of the Second Street Bridge was in the public's best interest, the city of Allegan developed a case to support the plan and to win federal funding. Under the leadership of City Manager Joanne Wrench, the city contracted for a structural study of the bridge and an analysis of the bridge's role in the city's transportation system. This study by a prominent engineering firm within the state became a powerful tool for the city, because it documented the structural integrity of the span and revealed that the city did not require anything more than a one-lane bridge at the site. Armed with these studies, and in the absence of any local interests actively promoting a new bridge, the city was able to marshal political support and persuade the state department of transportation that restoration of the bridge in a safe, yet historically sensitive, manner was appropriate. Following this, state and local officials urged the Federal Highway Administration to allow expenditure of federal funds on the project even though it would not meet AASHTO standards. Eventually the city received authorization from both state and federal highway departments for the restoration work. Interestingly, when the FHWA did approve the project in 1981, it stipulated that funding would come not from the bridge program, but from the smaller and less well-known Rural Secondary Road Fund.

Although this funding switch clouded the meaning of the FHWA's Allegan decision in relation to the expenditure of federal bridge funds, it did little to dampen the spirits of Allegan residents in the wake of their victory. Restoration of the Second Street Bridge cost $627,500 (versus an estimated $1.2 million for a new structure), and it is clear that the city considers preservation of the historic span an important concern. The project involved moving the entire bridge onto dry land and then disassembling all its members so their structural integrity could be checked. The structure was then repainted, reassembled and moved back to its original location, where it was reopened to traffic. The restoration involved the replacement of all vertical compression members with new steel members designed in a historically sensitive manner, but many other major iron components from the 1886 structure did not need to be replaced. In June 1983 the city celebrated the restoration of its historic span with a three-day civic bridge fest. Following the reopening, the century-old tradition of motorists' taking turns driving in opposite directions across the bridge continued, although the agreement for the refurbishing technically required that bridge traffic be one-way out of town. After some subsequent negotiations among federal, state and local officials, the city has now installed a traffic light system to control use of the bridge in two directions.

In retrospect, three important lessons can be learned from the Second Street Bridge story:

■ In bridge preservation cases, it is important to enlist the services of an engineering firm that will analyze a historic bridge's assets. In addition, traffic studies can be useful in documenting why a historic bridge will adequately meet a region's transportation needs.

■ Engineering studies can be used effectively to influence political decisions related to the approval of federal funds for bridge projects not meeting AASHTO standards. This is especially true when there is little local resistance to preserving a historic bridge.

■ Often, certain parts of 19th-century iron bridges can be in much better structural condition than their 20th-century steel counterparts, because wrought iron is much less susceptible to rust and corrosion than steel. For example, the iron pins used to connect the structural members of the Second Street Bridge were in excellent condition and were reused in the restored structure.

ORGANIZING SUPPORT FOR PRESERVATION

Structural and traffic engineering studies documenting the future utility of a historic bridge will not necessarily ensure its preservation. Despite such a study, the Bellows Falls Arch Bridge (1905) in Vermont was demolished in 1982 after the state and the Federal Highway Administration asserted in a section 4(f) report that no prudent or feasible alternatives to demolition existed. The validity of this conclusion was never put to a legal test. However, in a controversy involving the Pasco-Kennewick Bridge in eastern Washington, a similar assertion was successfully challenged in court by a dedicated group of local citizens.

■ **Pasco-Kennewick Bridge**
Between Pasco and
Kennewick, Wash.
Union Bridge Company
1922

The Pasco-Kennewick Bridge over the Columbia River, a cantilever through truss, is an excellent example of a long-span highway bridge designed during the early years of the automotive era that proved significant in the state's commercial development.

In the early 1970s, federal funds were provided to the cities of Pasco and Kennewick for construction of a replacement bridge, which was completed in 1978. Shortly thereafter, the old bridge was declared eligible for inclusion in the National Register, and the requirements of section 106 of the National Historic Preservation Act and section 4(f) of the Department of Transportation Act were triggered. A memorandum of agreement was executed in 1980 by the Advisory Council on Historic Preservation, the Washington state historic preservation office and the Federal Highway Administration. The agreement provided that the FHWA would not approve funds for demolition if a majority of the citizens of the two cities opposed demolition. If demolition were favored, then the secretary of transportation, acting through the FHWA, would conduct a review, pursuant to section 4(f), to determine if a feasible or prudent alternative existed. The citizens of both cities voted in favor of demolition. In

response to a draft 4(f) statement, the U.S. Department of the Interior recommended that the FHWA engage in concerted federal and local planning to develop an adaptive use plan for the old bridge. Preservationists proposed alternatives based on a feasibility study funded in part by the National Trust. In spite of these recommendations, the FHWA found none of the alternatives feasible or prudent. Preservationists then sought a remedy through the courts.

The U.S. District Court for the Eastern District of Washington upheld the 4(f) determination, apparently accepting without review the FHWA's determination. Preservationists appealed the decision to the U.S. Court of Appeals for the Ninth Circuit. To prevent demolition before the appeal could be heard, Virginia Devine, an active member of the local preservation group, personally posted most of the required $75,000 bond that extended the injunction prohibiting demolition. Due in large part to the extraordinary commitment of one individual, preservationists were able to marshal their efforts in the Ninth Circuit.

Top: Pasco-Kennewick Bridge, with the modern cable-stayed replacement span in the background. Above: Bridge preservation advocate Virginia Devine.

The Ninth Circuit held that the secretary of transportation acted arbitrarily in concluding that there were no feasible and prudent alternatives to demolition. The court found the 4(f) determination inadequate because it reviewed funding of alternatives only by the cities or the local preservation organization but failed to consider the availability of federal funds for rehabilitation or preservation. The decision should require the secretary to consider, in future situations, the use of federal funds both for rehabilitation of a historic bridge and for construction of a new bridge as a part of the same project. The court also affirmed conclusions reached in other 4(f) cases on the following issues:

■ The results of a ballot measure should not be the sole factor in the federal agency's evaluation. The court also pointed out that the meaning of the vote in this case was limited because the voters were not presented with sufficient options.
■ The cost of a rehabilitation alternative and community

disruption are factors to be considered, but, except for unusual circumstances, these should not suggest a particular result.

The Ninth Circuit ordered the case remanded to the district court for a new 4(f) determination by the secretary of the interior. As a result of this ruling, the bridge was granted a major reprieve and there is hope that a means can be found for preserving the historic span. While long-term preservation of the bridge remains in doubt, what is not doubtful is the power that bridge preservationists can wield in forcing the U.S. Department of Transportation to address the legal requirements of section 4(f). With passage of the most recent federal law concerning bridge rehabilitation and replacement, supporters of the Pasco-Kennewick Bridge have been given hope that federal funds, which would have been used for demolition, can be used for repairing and stabilizing the historic span.

■ ■ ■ ■ ■   DAM PRESERVATION   ■ ■ ■ ■ ■

Dams are considered such prosaic structures in most communities that they rarely attract the attention of historic preservationists. Even within the world of historians of technology and industrial archeologists, dams are not a subject of overwhelming interest. Consequently, there is no broad-based network that pays special attention to the preservation of historic dams. But dams play such an important role in the history of many areas that there is no reason why they should not warrant preservation interest. At the very least, people should be aware of some basic facts about how dams can fail and about legislative programs designed to ensure dam safety.

Before the 20th century dams were built by public agencies and private enterprises with no legal constraints. The only check on design was the fear that a structure's collapse could bring about the financial and professional ruin of engineers, business people and the public officials responsible for it. As part of Progressive era reforms in the early 20th century, several states initiated programs that placed dam design and construction under the legal supervision of state engineers or other state government authorities.

California, for example, passed a dam safety act in 1917 that gave the state engineer responsibility for approving the design and construction of all dams more than 25 feet high that were not built by a municipality (such as Los Angeles or San Francisco) or the federal government or fell under the supervision of the state railroad commission (predecessor of the public utility commission). In 1929 the law was revised so that all nonfederal dams in California came under the supervision of a special office in the state department of engineering. Today, this law is still in effect, with responsibility vested in the Division of Safety of Dams. The condition of dams in California is constantly monitored by this division and, as described in the following Littlerock Dam case study, it can take legal action to prevent storage of water behind a dam for perceived safety reasons.

Usually considered the most active state in the field of dam safety, California has served as a model for other jurisdictions. Dam safety is considered a state responsibility, with the federal government exerting control only over those dams that are federally owned. These primarily involve structures operated by the U.S. Bureau of Reclamation and the U.S. Army Corps of Engineers. Following the collapse of the Bureau of Reclamation's Teton Dam (1976) in Idaho in 1976, the federal government became concerned about the stability of the bureau's structures. Since that time it has authorized more than $700 million to inspect, analyze and rehabilitate dams under the agency's control. In 1986 the Water Resources Development Act authorized the Corps of Engineers to provide matching funds for states to inventory dams and develop permanent dam safety programs. In keeping with the tradition of separating state and federal authority in such matters, this legislation does not authorize the expenditure of federal funds to support the actual repair or alteration of nonfederal dams.

WHY DAMS FAIL

Under normal loading conditions dams are usually reliable structures that rarely fail. In fact, a study of dam failures shows that if a dam safely withstands the water pressures exerted when the reservoir is completely filled for the first time, it is unlikely that any future filling of the reservoir will precipitate collapse. For example, the Teton Dam (1976) in Idaho and the St. Francis Dam (1928) in southern California failed shortly after being subjected to the hydrostatic load of a full reservoir for the first time. In both of these cases it was not the dam itself that was improperly constructed. Rather, the rock foundations that the dams were built on were not sufficiently strong or impervious to support the load exerted by the reservoir.

Just because a dam survives its initial period of operation does not mean that it will never fail. Special circumstances, however, are usually required to bring about disaster. To one degree or another all dams leak, even if only a relatively small amount. Engineers become concerned only when the rate of leakage begins to accelerate, indicating that something is changing the overall equilibrium of the design. For example, a classic mode of failure starts with the growth of trees on top of an earthfill dam. At first it might be thought that the tree roots will help hold together the earthen structure. But in reality, the roots facilitate seepage through the dam by providing a path for the water to flow. As this seepage begins washing away small amounts of earth, it provides room for further flow. The erosion thus increases and eventually will cause the entire dam to wash away.

Most concern about dam safety focuses on overtop-

Bartlett Dam on the Verde River in Arizona. Surplus water in the reservoir is being released through the large spillway on the left.

ping and the maximum probable flood that can be expected to impact upon a dam. Overtopping occurs when more water flows into a reservoir than can be handled by spillways that divert water downstream around a dam. Many types of dams, especially those on large rivers, are designed to withstand overtopping, and these are often referred to as overflow dams. But other types, especially earthfill and rockfill structures, cannot survive any extended period of overtopping because they will erode and wash away. The famous 1889 Johnstown Flood in Pennsylvania was caused by overtopping of an earthfill dam during a heavy rainstorm.

Calculation of maximum probable floods involves statistical extrapolation of known rainfall data for a given region. Often these extrapolations are expressed in terms such as a "100-year flood" (a flood that can be expected to occur every 100 years) or a "1,000-year flood." Recently, the maximum probable flood calculations for various watersheds have been substantially increased. This means that spillways built to carry floodwaters around dams are often found to be inadequate and that repair work on many "unsafe" dams focuses almost exclusively on increasing their spillway capacity.

Earthquakes are often perceived as posing grave threats to the stability of dams. While a Hollywood filmmaker may view the idea of an earthquake causing a dam to breach, wiping out thousands of people, as a surefire way to capture an audience's attention, in reality practically no dams have collapsed as a result of seismic disturbances. In fact, to the author's knowledge no type of concrete or solid masonry dam has ever failed because of an earthquake. In contrast, earthfill dams, especially ones built using the hydraulic fill process, appear to be more vulnerable to seismic shocks. Hydraulic fill dams are formed by piling up large quantities of water-saturated earth and then allowing the excess water to drain away. The earthen structure left behind is then used to form the dam. The problem here is that the water used in the construction process never completely drains away, so the structure is left permanently soggy. When an earthquake shakes this type of dam, the earthfill can

Building hydraulic fill dams. Top: Directing pressurized water toward an earthen hillside. Left: Resulting "liquid earth" being channelled to the dam site and deposited via flumes and pipes. Below: Crane Valley Dam (1903) near Fresno, Calif., after the sluicing process has been completed. Excess water is slowly draining away.

easily turn into an unstable liquid mass (a process called liquefaction) and thus precipitate collapse. This is exactly what happened to the San Fernando Dam in Los Angeles when it partially failed during a 1971 earthquake.

Concern about overtopping from a maximum probable flood or collapse during an earthquake are the most common reasons why engineering studies recommend alteration or replacement of an old dam. In addition, dams may be abandoned or altered because they are perceived to be insufficient to meet modern water storage needs. Especially in the arid West, where water is such a precious commodity, the cost of building a new, larger dam to replace an existing structure can be justified even if the extra water storage capacity is relatively small. Unfortunately, raising the height of a dam is usually difficult unless it is drastically rebuilt or was originally designed to accommodate an increased height. Efforts to increase the storage capacity of a reservoir thus may result in an old dam being inundated by a new, larger dam built downstream. A good example of this is the 1884 Bear Valley Dam near San Bernardino, Calif., which was inundated by a taller dam completed in 1911.

DOCUMENTING COMMUNITY IMPORTANCE

Historic dams are not the typical focus of a preservation battle. Such confrontations usually occur when a community is faced with the loss of a historic house, a movie theater, a courthouse or perhaps even a bridge. But, as the example of the Littlerock Dam illustrates, historic dam preservation can become an issue of vital importance to a community.

■ **Littlerock Dam**
Littlerock, Calif.
John S. Eastwood
1924

Littlerock, Calif., is a small farming community in northern Los Angeles County. In 1892 a group of farmers formed the Littlerock Creek Irrigation District to develop pear and fruit orchards in land adjacent to Little Rock Creek (spelled differently from the town). This creek flows north out of the San Gabriel Mountains and carries a considerable flow from snowmelt in the spring. During the rest of the year, however, it is quite dry. From the beginning the farmers of Littlerock recognized the value of building a storage dam to impound the creek's spring floods and thus make more water available later in the year. Their problem was to raise sufficient funds to construct a large dam without forcing the district into bankruptcy.

Finally, in 1918 the district, along with its nearby partner, the Palmdale Irrigation District, arranged for noted engineer John S. Eastwood to design a multiple-arch dam that would be both affordable and tall enough to store most of the creek's floodwaters. In fact, it was to be the tallest multiple-arch dam in the United States at the time of its completion. Because of its height (more than 170 feet) the dam became controversial, and it took more than four years for the state engineer to approve the design. Other designs were explored but proved too

Above: Downstream side of the Littlerock Dam, with its spillway on the right. Left: Upstream face showing the multiple arches. Below left: Downstream face with buttresses and strut-tie beams near the angle in the dam. Below right: Formwork impressions in the concrete buttresses.

expensive for the two irrigation districts to afford. Simply put, either the districts were to build a multiple-arch dam or they could have no dam at all.

The districts received final approval for Eastwood's design in late 1922, and the dam was accepted as complete by the state engineer in June 1924. It soon began storing runoff from Little Rock Creek and proved to be a boon to the local economy. Although Littlerock never became a major agricultural center, it thrived as a

Above: Natural condition of arid desert land surrounding Littlerock. Above right: Orchards made possible through irrigation provided by the Littlerock Dam.

regionally significant area for pear production. The increased produce made possible by the dam was used to help pay off the construction bonds.

In terms of engineering, the structure also proved to be a great success. For more than 60 years the Littlerock Dam has provided exemplary service and fully lived up to the expectations of its designer. In fact, in early March 1938 its original spillway became clogged with debris during a heavy rainstorm, and water overtopped the dam by approximately two feet for several hours. Eastwood's design weathered this potential calamity without incident, and the dam suffered no damage from the unforeseen overtopping. The tallest of Eastwood's 17 multiple-arch dams, the Littlerock Dam is listed in the National Register of Historic Places.

Despite the success of the dam in weathering the 1938 flood, questions about the design remained in the minds of state dam safety authorities, perhaps a legacy of the original controversy. In the mid-1960s the Division of Safety of Dams ordered the two districts to analyze the structure's safety. Completed by a large engineering firm, this study arrived at the startling conclusion that the dam would be in danger of collapsing if the reservoir were allowed to fill with water. Using a simplified method of mathematical analysis, the study concluded that the upstream face of the dam's buttresses would be subject to dangerous "tensile stresses" if the reservoir were filled. The facts that the reservoir had been filled numerous times, including the overtopping of 1938, and that no "tensile cracks" had appeared on the upstream faces of the buttresses were ignored.

Buoyed by this seemingly incongruous study, state authorities began pressuring the two districts to fix their "unsafe" dam. The districts resisted, largely because of an innate faith in the dam's stability and because any repair work was beyond their financial capability. The situation remained in limbo for several years until the state made moves to revoke the permit that allowed the districts to store water behind the dam. To counter this, local residents formed the Citizens Committee to Save the Littlerock Dam to fight the state in court. Formation of this citizens group was necessary because the districts, having funded the engineering study that determined the dam to be unsafe, were parties to its conclusions.

In early 1977 the Division of Safety of Dams went into state court and, claiming that the dam represented an imminent threat to public safety, attempted to require that the reservoir be drained. As stated in court, the division's actions were based entirely on the engineering report prepared in the 1960s. The staff considered the report to be evidence of the structure's instability and felt they were acting only so that the districts would live up to their legal responsibilities and fix the "unsafe" Littlerock Dam. In response, the Citizens Committee engaged the services of Glenn L. Enke, a professor emeritus of civil engineering at Brigham Young University who had previously undertaken a study of Eastwood's Mountain Dell Dam (1917, 1925) for Salt Lake City. After examining and analyzing the Littlerock Dam, Enke testified that he could not agree with the conclusions of the earlier engineering study. In his judgment, the Littlerock Dam was a safe and stable structure that represented no threat to public safety.

Presented with divergent opinions on the safety of the dam, in April 1977 the court ruled that the state could not proceed to force immediate drainage of the reservoir. In finding that no imminent emergency existed, the court directed the state to follow due process, such as completion of an environmental impact report, before undertaking any future effort to revoke the dam's permit to store water.

Since the court ruling, the dam has continued to function without any design changes. The irrigation districts are still negotiating with the state over the status of the dam and what needs to be done to "fix" it. The conclusions of the 1960s engineering study have largely become passé — replaced by concern over the possible effect of an earthquake. Although concrete dams have a superb record of withstanding earthquakes, apprehension about the Littlerock Dam has been heightened because it is located only a few miles from the San Andreas Fault. The future of the historic structure remains uncertain as engineers and the state continue to analyze the structure.

RAISING ISSUES OF PERSPECTIVE

The ongoing battle to save the Littlerock Dam and preserve its role in the local agricultural economy has involved many concerns specifically related to the form and history of this particular dam. But some aspects of the controversy raise more general questions about the preservation of historic dams:

■ To demonstrate the potential disaster that could result from failure, the state stipulated a maximum probable flood almost three times the size of the largest known flood along Little Rock Creek. This flood was based on a rainstorm in which 29 inches of rain would fall over the entire 64-square-mile watershed above the dam within a period of 36 hours. Using this flood potential, and assuming the complete collapse of the dam, the state determined the area that would be inundated. Amazingly, the only major site in the inundation pool was a small trailer park through which a few feet of water would flow for a short period of time.

Many Littlerock residents expressed bewilderment that the state would spend so much time worrying about the Littlerock Dam when, even by the state's own calculations, so little existed in the floodplain below the dam. As far as the state is concerned, its interest in providing safe dams is justified even if only one person's life might be threatened by a faulty structure. But, according to the people of Littlerock, the state's position would bring economic devastation to the entire community to protect a few homes in a trailer park. The safest dam on Little Rock Creek would, of course, be no dam at all, but this would mean the wasting of valuable floodwaters and the loss of many existing orchards.

■ The Littlerock case highlights how maximum probable floods can be based on extreme meteorological conditions. The idea that 29 inches of rain will fall in a period of 36 hours over a 64-square-mile area in southern California is something that even Noah would have found rather startling. Certainly, the region is susceptible to infrequent heavy flooding, but this hypothesis strains the bounds of credulity.

■ The Littlerock Dam controversy points out the vagaries possible in "objective" engineering studies. Depending on the assumptions used in a mathematical analysis, a wide spectrum of conclusions about a structure's safety is possible. Clearly, any group seeking to dispute an engineering study about a historic dam must seek out an engineer who can analyze the structure from a fresh and knowledgeable perspective. The structure may, indeed, be in a dangerous condition, but this should not be assumed just because one study indicates it to be so. The Citizens Committee to Save the Littlerock Dam gained considerable credibility by hiring an expert on multiple-arch dams to undertake an independent study. Without Professor Enke's study and his court testimony, the people of Littlerock would have had much more difficulty countering the state's assertions.

■ Efforts to preserve the Littlerock Dam illustrate the value of bringing public attention to all the ramifications of the issues. Rather than let the situation be portrayed merely as a case of an "unsafe" dam threatening residents, it is important that the dam's fate be discussed in terms of the economic hardship that would result if water could no longer be stored behind it. Under extreme circumstances the dam might pose a threat to a few people below the structure. But its forced abandonment would unquestionably result in great economic hardship for the area as a whole. Controversies such as the Littlerock Dam case should be presented not just as an antiseptic problem of dam safety, but as an issue relating to the very survival of a community. Once this has been established, then technical arguments over such things as mathematical assumptions and "tensile stresses" can be placed in a broader perspective.

■ ■ ■ ■ INNOVATION AND FLEXIBILITY ■ ■ ■ ■

The preceding examples provide an overview of the numerous issues involved in preserving bridges and dams. New and different concerns arise continually. For example, in Washington, D.C., the effect of suicide barriers on the visual and architectural character of the Duke Ellington Bridge (1931) across Rock Creek Park has been debated in the courts. The question of how to balance the aesthetic loss from the addition of barriers with the sought-after societal good is not easy to answer. And, although dams are not nearly as endangered as bridges, a proposal has recently been aired to tear down the O'Shaughnessy (Hetch Hetchy) Dam (1923, 1938) in California's Yosemite National Park and reclaim the valley that was inundated after one of the country's earliest environmental skirmishes.

Bridge and dam preservationists have had to be innovative, arranging in some instances to store historic spans for later use on another site. Other bridges have been restricted in the type of traffic they carry, being changed to one-way traffic or pedestrian or bicycle use. While the solutions vary, preservation efforts have been most successful when they have been based on well-developed and well-informed public support. The importance of this support in an ongoing effort to save these sites is evident in cities such as Pasadena, Calif., where the local private, nonprofit preservation group and the city are working together in a long-term effort to preserve the Colorado Street Bridge (1913). Many historic bridges and dams already have been lost, and more will be demolished if public awareness and local support are not strong and are not used effectively to advocate the advantages of preservation and sensitive rehabilitation.

Steel suicide barriers recently installed on the deck of the Duke Ellington Bridge (1931) across Rock Creek Park in Washington, D.C.

Note: Some bridge information and examples here have been taken from *Saving Historic Bridges*, Information Series no. 36, prepared by Donald C. Jackson, Nancy C. Shanahan, Elizabeth Sillin and Vincent Marsh (1984, National Trust for Historic Preservation).

GUIDE TO THE GUIDE

In selecting sites for inclusion in this book, care was taken to find bridges and dams that still exist and are relatively accessible to the public, so readers can actually go and examine them firsthand. Many historically significant structures, especially those dating to the early and mid-19th century, have long since been demolished and thus are not included. A number of interesting and unique bridges and dams survive on private property, but they are not generally open for the public to visit or view and hence are not included here. Of the listed sites, some may be seen from a slight distance, but entry to the structures themselves may be prohibited by the companies or public agencies that control them. Readers are counseled to obey all laws governing access to the bridges and dams described in this book. As with any type of engineering or industrial site, bridges and dams can present hazards to visitors (and even trained experts) and thus should be avoided if any questions exist as to safety.

Beyond attempting to ensure public accessibility, several other factors were considered in selecting sites. Because this book is published by the Preservation Press of the National Trust for Historic Preservation, emphasis has been placed on structures that are at least 50 years old. A few major sites completed in the 1940s, 1950s and 1960s are also included, but, in general, the majority date from the late 19th to early 20th centuries, because of the intended focus on America's earlier technological heritage.

And, in addition to all the great American bridges and dams — the large, well-known landmarks that contributed to the growth of our important cities and outlying areas — care has been taken to direct attention to small structures that survive as representative examples of once-common technologies. This means that along with widely known urban bridges, the book includes descriptions of more modest rural spans that, at the time of their construction, no one would have considered "great" American bridges. But so many of these small bridges have been replaced or demolished in recent years that the few surviving ones have taken on greater significance as bridge types that were extremely important in America's economic development. For this reason they are included in this guide.

A final criterion for selection concerned geographical distribution: sites from all parts of the United States have been included. Obviously, some states have more sites and are thus better represented than others. Because dams play a more important social and economic role in the arid West, western dams receive more emphasis than their eastern counterparts. In some instances a structure located in a less populated or more recently settled area is included that would not be considered noteworthy if found elsewhere. Rather than judge sites against a single, national standard of significance, an attempt has been made to assess their importance within more localized contexts. The intent is to provide a full range of the various types and sizes of bridges and dams that contributed to America's development, for our nation's technological heritage constitutes much more than just a few structures that can be labeled "first" or "biggest." To readers who find no mention of one of their favorite structures the author apologizes; that a bridge or dam is excluded from the book does not mean it lacks historical significance.

The bridge and dam entries in the following guide section are organized into six regions, beginning with the Northeast and moving south and then westward. Within each region, the states, cities and sites are presented alphabetically. Locations indicate the nearest named town or vicinity and generally include route numbers as well as natural features spanned (a river or valley, for example). Because many of the sites are in remote areas or small towns outside larger metropolitan areas, travelers first may want to review the index to find the bridges and dams nearest the more well-known cities they plan to visit.

Each entry highlights key details at the left or right. First to appear is the name or names by which the site has been known, with any secondary name in parentheses. Below this appear the natural feature spanned and the route number or road name. The next line gives the engineer, chief engineer, fabricating company, construction agency or architect; local builders are so designated. For a few sites, this information is unknown. The last line indicates the year in which the bridge or dam was completed; for bridges reerected on another site or substantially reconstructed, this year also is provided. In the entries themselves, "NR" or "NR district" indicates that the site or its adjacent district is listed in the National Register of Historic Places individually or as part of a historic district, and "ASCE" designates a National Historic Civil Engineering Landmark of the American Society of Civil Engineers. A final note: dimensions given for structures in some instances have been rounded off to the nearest foot. In addition, different sources often give conflicting dimensions for certain structures. In particular, the height of dams, which is measured from deepest foundation to the crest, can be difficult to determine with absolute certainty.

State Street Bridge, Chicago (demolished), a two-leaf bascule bridge across the Chicago River in an open position, c. 1905.

NEW ENGLAND

Interior view of the Cornish-Windsor Bridge (1866), a Town lattice truss across the Connecticut River between New Hampshire and Vermont.

■ ■ ■ ■ ■   CONNECTICUT   ■ ■ ■ ■ ■

BURLINGTON

■ **Nepaug Dam**
Across the Farmington River
West of State Route 179
Caleb M. Saville
1914

In 1914 the Hartford Board of Water Commissioners began construction of a new water supply system to provide additional water to the state capital. For many years the city had eyed the Farmington River as a potential source of water, but plans to dam a tributary, the Nepaug River, were stymied by downstream mill owners dependent on the river for water power. Finally in 1911 the impasse was resolved when the city agreed to build a "compensating reservoir" in the headwaters of the Farmington River to store floodwaters for later use by the mill owners. With this political accommodation the state legislature authorized construction of several dams to impound the Nepaug River for municipal use. The largest of these was the Nepaug Dam, a concrete, curved gravity dam 113 feet high and 100 feet wide at the base. The ample dimensions of the structure were apparently justified by the placement of the overflow spillway in the center of the dam. The structure still functions as an important component of Hartford's water supply system, and it stands as testimony to the efforts made to accommodate growing municipal water needs and existing manufacturing interests in the early 20th century.

COLCHESTER

■ **Lyman Viaduct**
Across Dickinson Creek
⅕ mile west of
Bull Hill Road
Boston and New York
Airline Railroad
1873, 1913

Early railroad service between New York City and Boston followed a route parallel to the Connecticut–Rhode Island coastline until it approached Narragansett Bay. At that point the right-of-way turned north and headed to Boston via the city of Providence. Although not the most direct, this route allowed the railroads to avoid crossing many of the small, yet deep, valleys that run north and south through eastern Connecticut. In the early 1870s the Boston and New York Airline Railroad sought to reduce the transit time between the two cities by building an "air line" route that cut across these valleys. The most imposing of the valley crossings built along this route is the 1,100-foot-long Lyman Viaduct. Rising 137 feet above Dickinson Creek, the viaduct consists of a wrought-iron trestle formed using eight-inch diameter Phoenix columns. The viaduct came under the control of the New York, New Haven and Hartford Railroad in 1882, and it remained in service without major alteration for almost 30 more years. At that time the company became concerned that heavy train loads were threatening the safety of the structure, and it filled in the trestle with fine sand covered by a layer of compacted cinders. In essence, the wrought-iron trestle became a kind of skeleton buried inside a huge earth embankment. The viaduct no longer carries railroad traffic, but it was recently adapted to carry a sewer line for the town of Colchester across the top of the structure. This new use had a minimal impact on the integrity of the 1873 trestle buried in the embankment. NR.

EAST MORRIS

Between 1893 and 1901 the city of Waterbury constructed the Wigwam Dam, an all-masonry gravity dam built along a slight curve, as the key component of its new water supply system. Upon its completion the 91-foot-high dam was one of the tallest water impoundment structures in New England. The interior core is rubble masonry, while the upstream and downstream faces consist of random ashlar granite from Connecticut and Vermont quarries. Smaller than other contemporary masonry gravity dams — the Wachusett Dam (1905), near Boston, and the New Croton Dam (1907), near New York City—the Wigwam Dam is nonetheless an excellent example of a large-scale curved gravity dam built in the preconcrete era.

■ **Wigwam Dam**
Across the West Branch of
the Naugatuck River
1 mile south of State
Route 109
R. A. Cairns
1901

Downstream side of the all-masonry Wigwam Dam.

GREENWICH

Still used to carry vehicular traffic over the busy Northeast Corridor railroad right-of-way, the Riverside Avenue Bridge is among the few active highway bridges in the United States to use substantial amounts of cast iron. As documented by Matt Roth in a recent inventory of historic engineering and industrial sites in Connecticut, the double-intersection Pratt structure was originally part of an 1871 six-span railroad bridge over the

■ **Riverside Avenue Bridge**
Across the Northeast
Corridor railroad line
On Riverside Avenue
Francis C. Lowthorp
1871; reerected 1890s

Left: Riverside Avenue Bridge. Below: Detail showing the diagonal rods through the vertical compression members.

Housatonic River in Stratford, Conn. In the 1890s a single span from this bridge was reerected on Riverside Avenue in Greenwich. Although the structure's floor system dates to the 20th century, as a whole the iron truss with its cast-iron compression members and wrought-iron tension rods retains a remarkable amount of historical integrity. Aside from cast-iron "gingerbread" that festoons the design, the age of the structure is evident in its use of vertical end posts. By the 1880s essentially all Pratt trusses built in America used inclined end posts because they were more economical. The 164-foot-long bridge is also one of the only surviving double-intersection Pratt trusses in which the diagonal tension members pass directly through rather than to the side of the vertical compression members. The structure's ability to function as part of a modern road system can be traced to its original use as a railroad bridge. Because it was designed to carry relatively heavy locomotives and rolling stock, it still safely supports vehicular traffic. NR.

NEW MILFORD

■ **Boardman's Lenticular Bridge**
Across the Housatonic River
On Boardman Road
Berlin Iron Bridge Company
1888

During the late 19th century the Berlin Iron Bridge Company became one of New England's most important structural iron fabricating firms. Based in Berlin, Conn., the company specialized in the construction of lenticular truss bridges based on the patents obtained in 1878 and 1885 by William O. Douglas. The lenticular truss is distinguished by polygonal top and bottom chords that form a lens shape. Although this design feature required more complicated shop work, it also reduced the amount of metal needed for spans of any given length compared to standard Pratt trusses. According to bridge historian Victor Darnell, the firm built several hundred of these patented spans in the 1880s and 1890s before being absorbed into the Andrew Carnegie–controlled American Bridge Company in 1900. In 1889 the company claimed to have built 90 percent of the iron bridges erected in New York and New England during the

Boardman's Lenticular Bridge before it was closed to traffic. The ornate portal is characteristic of many Berlin Iron Bridge Company structures.

previous 10 years. Although this claim is certainly an exaggeration, the company was extremely prolific, and its agents marketed lenticular trusses as far away as Texas.

The Boardman's Lenticular Bridge spans 188 feet and is the longest lenticular truss surviving in Connecticut. Recently, the structure was bypassed by a new highway bridge built adjacent to the historic span. It stands today as a major example of a highly distinctive truss type that made significant contributions to New England's economic development. NR.

Lens shape of the Boardman's truss design.

■ ■ ■ ■ ■ ■ ■　MAINE　■ ■ ■ ■ ■ ■ ■

BAILEY ISLAND

Bailey Island lies a short distance off the coast of central Maine. In the 1920s the Maine Department of Highways sought to connect the island and the mainland with a permanent bridge. Because of the harsh marine environment and the steady erosive effect of tidal flow in Casco Bay, however, no standard bridge design appeared adequate. To meet the unique challenge posed by the site, in the early 1920s the state bridge engineer, Llewellyn N. Edwards, developed a special design — a 1,120-foot-long granite bridge constructed so that large openings between the masonry slabs allow water to flow through the structure in response to tidal variation. In addition, a

■ **Bailey Island Bridge**
Across Casco Bay
On State Route 24
Llewellyn N. Edwards
1928

Deck of the Bailey Island Bridge, built on masonry slabs.

Detail of the stone crib design that allows water to pass through the Bailey Island Bridge.

short-span, reinforced-concrete girder in the middle of the bridge permits passage of small boats. The rough-hewn masonry is not susceptible to rotting or rusting, and the bridge does not suffer from the maintenance problems that would affect a wooden or steel structure. After more than 50 years of service, it is still in good condition and in active use. NR, ASCE.

ELLSWORTH FALLS

■ Ellsworth Dam
Across the Union River
Adjacent to U.S. Route 1A
Ambursen Hydraulic
Construction Company
1907

Left: Penstocks leading to the Ellsworth Dam's hydroelectric powerhouse. Right: Downstream face of the overflow spillway.

This structure on Maine's north coast is one of the earliest large-scale examples of an Ambursen flat-slab buttress dam. This type of reinforced-concrete structure has a flat upstream face that is built on an angle of approximately 45 degrees and is supported by a series of reinforced-concrete buttress walls that, in the case of the Ellsworth Dam, are spaced 18 feet apart. Each of these buttresses rests directly on the bedrock foundations. Built for the Bar Harbor and Union River Power Company, the Ellsworth Dam is 65 feet high and 450 feet long, 275 feet of which are designed as an overflow spillway. The power plant is located within a separate building directly downstream from the dam, but much of the space between the buttresses was originally used as storage rooms and workshops. After more than 80 years of reliable service, it is still a productive part of the regional electric power system. NR.

Ellsworth Dam and
powerhouse.

NEW PORTLAND

As one might expect, the wire noted in this bridge's name refers to the form of its design. Completed shortly after the Civil War, it is a rare surviving example of a suspension bridge with shingle-covered wooden support towers. The structure has a clear span between towers of 198 feet and extends 30 feet above the river's normal water level. The towers rise a little more than 23 feet above the roadway. In the early 1960s the state highway commission carefully renovated the structure and took care not to alter its original form. Part of this work involved stabilizing the foundation of the west tower with concrete to make sure that it cannot slip into the river channel and cause the structure to collapse. Located on a little-used road, the New Portland Wire Bridge still carries local highway traffic. NR.

■ **New Portland Wire Bridge**
Across the Carrabasset River
On Wire Bridge Road
c. 1868

Left: Shingled wooden
towers of the New Portland
Wire Bridge. Above: Detail
of the wire cable connection
and anchorage.

Paddleford truss design used
for the Lovejoy Bridge.

SOUTH ANDOVER

■ **Lovejoy Bridge**
Across the Ellis River
¼ mile east of State Route 5
1868

This covered bridge is a relatively rare Paddleford truss, a
type of covered bridge that achieved a small degree of
popularity in New England during the mid-19th century.
The truss resembles the Town lattice truss in its use of
rigidly connected, intersecting diagonal members, but it
required many fewer members than the Town truss and is
a much lighter structure. This truss type was built
primarily in rural areas to carry light wagon loads. The
Lovejoy Bridge is still used for light traffic. NR.

VERONA

■ **Waldo-Hancock Bridge**
Across the Penobscot River
On U.S. Route 1
David B. Steinman
1931

Named after the two counties it joins, the Waldo-
Hancock Bridge was the first permanent bridge across
the Penobscot River below Bangor, located more than 20
miles upstream. Because of the importance of navigation
along the lower Penobscot River, a long-span design was
needed to provide adequate clearance for ships. The
design commission went to Robinson and Steinman, a
New York–based engineering firm that specialized in
suspension bridges. Designed under the direction of
David B. Steinman (1886–1960), the structure is 2,040

Left: Roadway and steel
tower of the Waldo-Hancock
Bridge. Right: Bridge rising
above the Penobscot River.

feet long and has a clear span of 800 feet between towers.
The deck is 135 feet above water level, a distance more
than enough to allow the passage of large ships up the
river. Completed in 1931, the bridge is still in use as part
of a relatively heavily traveled section of Route 1. NR.

■ ■ ■ ■ ■ ■ MASSACHUSETTS ■ ■ ■ ■ ■ ■

CANTON

The Boston and Providence Railroad, perhaps the most important of New England's early lines, connected two of the area's most important commercial centers and helped demonstrate the financial feasibility of large-scale railroad operations in a region until then dependent largely on maritime transportation systems. Completed in 1835, the Canton Viaduct was the largest and most significant structure on the railroad's main line. The East Branch of the Neponset River is a fairly small, sluggish stream, but crossing it on a level grade required a 615-foot-long masonry structure that extends 70 feet above the stream's normal level. Water in the river passes through six arches, each with a span of about eight feet. A 22-foot-long arch also provides clearance for a road through the viaduct. The bridge appears to be a solid mass of masonry, but it is actually a hollow structure with a distance of nine feet separating the five-foot-thick outer walls.

■ **Canton Viaduct**
Across the East Branch
of the Neponset River
Near Neponset Street
William G. McNeill
1835

CLINTON

In 1895 the Massachusetts State Board of Health issued a report recommending a major expansion of the water supply system serving the metropolitan Boston area. Faced with a rapidly expanding population, Boston needed a new source of uncontaminated water to stave off a potential economic and sanitary crisis. Under the leadership of Frederic P. Stearns, the board studied numerous rivers and lakes in Massachusetts and even some in New Hampshire before selecting the South Branch of the Nashua River as the best watercourse to impound with a new storage dam. Located more than 40 miles west of Boston, the dam and reservoir were to deliver water through a 12-mile tunnel emptying into the Sudbury Reservoir, then being built by the city of Boston. From there, water from the Nashua River could be distributed to the Boston region. In 1895 the state legislature authorized formation of the Metropolitan Water Board to build the Wachusett Dam. Stearns became chief engineer for the new board and, with advice

■ **Wachusett Dam**
Across the South Branch
of the Nashua River
Adjacent to State Route 70
Frederic P. Stearns
1905

Downstream face of the all-masonry Wachusett Dam, which provides water for the greater Boston area.

from John R. Freeman (1855–1932), supervised design and construction of the straight-crested masonry gravity dam. The Wachusett Dam is 944 feet long with an overflow spillway more than 450 feet in length; it has a maximum height of 205 feet. It remains in use as an active component of the Boston region's water supply system.·

HOLYOKE

■ **Holyoke Dam**
Across the Connecticut River
Downstream from
U.S. Route 202
Clemens Herschel and
E. S. Waters
1899

Incorporated as a city in 1850, Holyoke became a major center of America's paper manufacturing industry. Construction of a large timber crib dam at Hadley Falls in 1849 diverted water into an extensive series of power canals in Holyoke and created America's largest 19th-century hydropower development. The timber crib dam functioned successfully for almost 50 years before being replaced by a masonry gravity dam. The new dam is an overflow structure designed with a smooth curve (known to engineers as an "ogee" curve) on the downstream face, a design feature that helps dissipate the enormous kinetic energy released by the water passing over the dam. The Holyoke power canals are no longer in use, but water diverted by the dam is still used to generate hydroelectric power.

Left: Holyoke Dam. Right:
Ponakin Road Bridge.

LANCASTER

■ **Ponakin Road Bridge**
Across the North Branch
of the Nashua River
On Ponakin Road
Watson Manufacturing
Company
1871

The Ponakin Road Bridge is a rare surviving example of the Post truss, developed by Simeon S. Post (1805–72) in the 1860s and used for American highway and railroad bridges before the 1880s. It is distinguished by wrought-iron compression members that are slightly inclined toward the center of the span. This feature supposedly reduced the amount of material required compared to a Pratt truss of similar dimensions. Like the double-intersection Pratt truss, the Post truss employs diagonals that extend across two panels. It was used for several large railroad structures in the 1870s, including the Union Pacific Bridge across the Missouri River at Omaha, and was well known among late 19th-century engineers. The design is often referred to as the Post patent truss, but, although Simeon Post did take out patents in the 1860s, research in U.S. patent records does not reveal that he ever patented a truss design featuring inclined compression members. The Ponakin Road Bridge is 100 feet long and provided service for more than 100 years before being closed to vehicular traffic in the late 1970s. NR.

LAWRENCE

After the successful hydropower development in Lowell, Mass., in the 1820s and 1830s, industrialists sought to find new sites capable of supporting large textile factories. About 12 miles downstream from Lowell on the Merrimack River lay Bodwell's Falls, a series of small rapids that offered opportunities for developing a large-scale hydropower project. The project, however, required construction of a large overflow dam capable of raising the river level more than 30 feet. In 1845 the Essex Company began construction of a 943-foot-long, 32-foot-tall, straight-crested masonry gravity dam on the site's bedrock foundation. Completed in 1848, it was the largest dam in the United States for many years and helped provide water power for the huge textile mills in Lawrence. Today, the dam is still in service and supplies water for a newly built hydroelectric power plant. NR.

■ **Great Stone Dam**
Across the Merrimack River
Near Broadway Street
Charles Storrow
1848

Left: Water spilling over the Great Stone Dam. Below: Intake from the dam to the Lawrence power canal system.

Pawtucket Dam, with its irregular alignment across the Merrimack River.

LOWELL

■ Pawtucket Dam
Across the Merrimack River
West of School Street
James B. Francis
1875

The city of Lowell is among the most famous industrial communities in the United States. Located at a bend in the Merrimack River, Lowell was built to capture the enormous hydraulic power available at Pawtucket Falls. In the 1820s a group of Boston capitalists began converting a transportation canal around the falls into a power canal, and this soon spawned a huge conglomeration of textile mills. The system relied on the diversion of water from the river into the power canals, a task fulfilled by the Pawtucket Dam. First built as a wooden crib dam in 1826 and subsequently replaced in 1833 and 1847, the existing Pawtucket Dam was entirely reconstructed in 1875 as a masonry gravity overflow dam with a maximum height of 15 feet and a length of 1,093 feet. In addition, its height can be raised a few feet by the placement of wooden flashboards on the top of the structure. These flashboards allow storage of more water behind the dam during periods of low stream flow. ASCE.

NEWTON

■ Echo Bridge
Across the Charles River
At Upper Newton Falls near
Ellis Street
Alphonse Fteley
1878

In the 1870s Boston expanded its water supply system and built an 18-mile-long aqueduct connecting the Chestnut Hill Reservoir with the Sudbury River west of the city. The aqueduct traverses relatively gentle terrain,

Echo Bridge, an aqueduct in a bucolic setting.

but the Charles River formed an imposing natural barrier along the right-of-way. To cross the river the aqueduct's chief engineer, Alphonse Fteley (1837–1903), designed a seven-span, stone arch bridge with a total length of 475 feet. Six of these spans are less than 40 feet long, but the main arch is 130 feet long. When built, it was the second-longest stone arch span in the United States after the Cabin John Aqueduct (1864) in Bethesda, Md. Its name derives from an acoustical phenomenon that is quite pronounced along the underside of the main arch. The structure still carries water for the Metropolitan Water District. NR.

NORTHAMPTON

The "American system" of pin connections for metal truss bridges allowed for both rapid construction on site and general assurance that the design would act as a statically determinant structure — that is, one capable of relatively simple stress analysis. Two drawbacks of pin-connected designs were that they could loosen under constant, heavy moving loads and that they were susceptible to excessive deflection at midspan. In the late 19th century several railroads began building metal lattice trusses with riveted connections to avoid these problems. Similar in theory to the wooden Town lattice truss, metal lattice trusses were much stiffer than pin-connected trusses and provided smoother passage for fast-moving trains. The multiplicity of riveted connections, however,

■ **Northampton Lattice Truss Bridge**
Across the Connecticut River
Adjacent to State Route 9
R. F. Hawkins
1887

Northampton Lattice Truss Bridge after the rails had been removed from the deck.

created a statically indeterminant structure, one that could not be analyzed for stress using simple mathematical techniques. Metal lattice trusses were rarely used for highway bridges, but, as with the case of the Boston and Maine Railroad's crossing of the Connecticut River in Northampton, railroads frequently chose to build them in spite of the expense engendered by riveting numerous connections in the field during the erection process. The Northampton bridge consists of six spans, each with a length of approximately 200 feet. The bridge is no longer used to carry railroad traffic, although proposals have been made to adapt it for pedestrian and bicycle use.

WARE

■ **Winsor Dam**
Across the Swift River
On State Route 9
Frank E. Winsor
1940

The expansion of Boston's water supply system into the Clinton River watershed in the early 20th century did not quench the metropolitan region's collective thirst. By the 1920s the Metropolitan District Commission sought to augment its existing system and cast its eyes farther westward to the watershed of the Connecticut River. As worked out in protracted legislative maneuvering, the key to the plan involved building a huge dam to impound the North Swift River in the central part of the state. A 26-mile-long tunnel would then connect the new reservoir with the existing Wachusett Reservoir and provide a means of reaching water users in the Boston area. From an engineering point of view, the plans for the new reservoir and aqueduct were quite elegant. But to residents in the Swift River Valley, the scheme threatened to destroy their way of life; this was especially true for people living in the towns of Enfield, Dana, Prescott and Greenwich, because the proposed reservoir would completely inundate their homes. With passage of the Swift River Act by the Massachusetts legislature in 1927 and subsequent approval by the U.S. Supreme Court, the objections of local residents were overcome and engineering planning began in earnest. Although some might have sympathized with the plight of the few thousand people in Enfield, Dana, Prescott and Greenwich, the physical needs (not to mention political clout) of hundreds of thousands of people in eastern Massachusetts were not to be denied.

Under the direction of Frank E. Winsor (1870–1939), an engineer with experience in building the Wachusett Dam (1905) and the Gainer (Scituate) Dam serving Providence, R.I., construction of the main dam began in the mid-1930s. Formally dedicated as the Winsor Dam in 1940 shortly following the chief engineer's death, the structure is a 295-foot-high earthfill dam with a crest length of more than 2,600 feet. To help reduce leakage, the structure has a large concrete cove wall that is tightly keyed into the bedrock foundation. It also has a rockfill covering on the upstream face. The Winsor Dam forms the Quabbin Reservoir, a huge man-made lake with a surface area of more than 38 square miles and a capacity of 412 billion gallons, or 1,265,000 acre-feet (an acre-foot equals a volume of water one foot deep covering an area of one acre). Clearing for the reservoir required destruction of the towns to be inundated, and in the spring of 1938, these communities were officially abolished. Shortly afterward the reservoir began to fill up with water. Present-day visitors are welcome to visit the dam. Because the reservoir remains in use as a vital part of the Metropolitan District Commission water supply system, for health reasons, access to the water itself is strictly controlled.

■ ■ ■ ■ ■   NEW HAMPSHIRE   ■ ■ ■ ■ ■

BRADFORD

The Bement Covered Bridge is a 63-foot-span Long truss built on a bridge site first used in the 1790s. Stephen H. Long, born in New Hampshire in 1784, became a prominent member of the U.S. Army Topographical Engineers, the predecessor of the U.S. Army Corps of Engineers, in the early 19th century. Among his more notable accomplishments was the patenting of a wooden truss bridge design in 1830. As described by covered bridge historian Richard S. Allen, the Long truss "resembled a series of giant boxed X's" and did not use any auxiliary arches to strengthen the span. In addition, it is an all-wood design. Long actively promoted his truss type and with the assistance of his brother, Moses, found considerable success in convincing town commissioners to erect his design. Although the Bement Bridge was extensively refurbished in the late 1960s, the structure remains a good example of the Long truss design and is open to light vehicular traffic. NR.

■ **Bement Covered Bridge**
Across the West Branch of
the Warner River
On Center Road
1854

South portal of the Bement
Covered Bridge, a simple yet
attractive design.

CORNISH

■ **Cornish-Windsor Bridge**
Across the
Connecticut River
On Vermont State Route 44
James Tasker and
Bela Fletcher
1866

The Connecticut River forms the boundary between New Hampshire and Vermont, and until the 20th century it presented a considerable barrier to commerce. In the early 19th century numerous toll bridges were built in New England, including several covered wooden bridges across the Connecticut. According to Richard S. Allen, the last wooden toll bridge to operate between New Hampshire and Vermont was the Cornish-Windsor Bridge. Built by two carpenter-engineers from nearby New Hampshire, it is a two-span Town lattice truss with a total length of 460 feet. Each span is only 203 feet long, however, as the lattice work extends for a considerable distance beyond the edge of the abutments. In terms of total length, it stands as the third-longest surviving covered bridge in the United States, after the Bridgeport Bridge (1862) in Grass Valley, Calif., and the Old Blenheim Bridge (1855), in North Blenheim, N.Y. Tolls were collected for more than 75 years before the state of New Hampshire assumed ownership in 1943. The bridge was closed to traffic in 1987 and awaits restoration. NR, ASCE.

Two-span Cornish-Windsor Bridge, one of the longest covered bridges surviving in the United States.

TILTON

■ **Tilton Island**
Park Bridge
Across the Winnipesaukee
River
In Tilton-Island Park
A. D. Briggs and Company
1881

Charles E. Tilton accrued a considerable fortune as a merchant and industrialist in the mid-19th century, and, possessed of a philanthropic bent, he subsequently lavished some of this money on his hometown. Among his more lasting gifts to the community is a park on Tilton Island. The park is in the middle of the Winnipesaukee River with access provided by this 83-foot-long, two-span, wrought-iron Truesdell truss bridge. The Truesdell truss is one of those odd truss designs that never attracted much interest among the engineering profession but occur just often enough that bridge historians cannot dismiss them as pure fancy or chicanery. Patented by Lucius Truesdell of Warren, Mass., in 1858, the

Truesdell truss used for the Tilton Island Park Bridge.

Truesdell truss is a small, visually attractive lattice structure designed to be built entirely of wrought iron. Exactly why this type of design was resurrected for use at Tilton Island Park in the early 1880s is unknown. Regardless, it remains in use today, as it has for more than 100 years, providing pedestrian access across part of the Winnepesaukee River. NR.

WEST OSSIPEE

Built near the New Hampshire–Maine border, the Whittier Bridge is a well-preserved example of a wooden Paddleford truss with an auxiliary arch. Although the present structure dates to shortly after the Civil War, it is located on a site apparently used for a bridge crossing as early as 1796. In 1869 heavy flooding washed away many bridges in the Ossipee region, and it is believed that the existing structure was built shortly afterward. While the facts are uncertain, the builder most likely was either Jacob E. Berry or his son, Jacob H. Berry. Whether or not the arches were part of the original design is also unclear. In any case, they are not of recent origin and would have been built within a few years of the original construction. With the relocation of State Route 25 in the 1940s, the Whittier Bridge was no longer required to carry heavy traffic and today it serves only local transportation needs. The structure maintains much of its original integrity, but the abutments have been strengthened with reinforced concrete. NR.

■ **Whittier Bridge**
Across the Bear Camp River
Adjacent to State Route 25
Jacob E. Berry or
Jacob H. Berry, builder
c. 1870

Whittier Bridge, with its partially exposed auxiliary arch. The arch provides additional strength.

One of New Hampshire's oldest covered bridges, the 1832 Town lattice truss in West Swanzey.

WEST SWANZEY

■ **West Swanzey Covered Bridge**
Across the Ashuelot River
On Main Street
Zadoc Taft, builder
1832

Among the oldest surviving examples of covered-bridge technology in New Hampshire is this two-span, 159-foot-long Town lattice truss in West Swanzey. The bridge was built in 1832 by Zadoc Taft, a local master carpenter. The town of West Swanzey has kept the structure in good condition, substantially renovating it in 1859, 1888 and, most recently, 1973. The latest work involved extensive repair of the roof and floor system and also entailed replacing many of the wooden pins with steel bolts. The bridge is still open to light vehicular traffic. NR.

WINCHESTER

■ **Ashuelot Covered Bridge**
Across the Ashuelot River
On Bolton Road near State Route 119
1864

New Hampshire is a heavily wooded state, so it is not surprising that many covered bridges were built here in the 19th century. This structure is an excellent example of a Town lattice truss bridge. Patented by Ithiel Town (1784–1844) in 1820, this truss type was used extensively in the northeastern United States throughout much of the 19th century. Although the design requires considerable amounts of wood and entails boring hundreds of holes in the structural members for even a modest-length span, it could be readily built by any patient carpenter. The Ashuelot Bridge is a two-span structure with a total length of 178 feet. Trucks are no longer permitted to cross it, but the bridge is still open to automobile traffic. NR.

■ ■ ■ ■ ■ ■ RHODE ISLAND ■ ■ ■ ■ ■ ■

BRISTOL

■ **Mount Hope Bridge**
Over Mount Hope Bay
On State Route 114
David B. Steinman
1929

Founded by Roger Williams in the mid-17th century, Rhode Island is centered around the expansive Narragansett Bay. Historically, much intrastate commerce was carried by ship and ferry. With the growth of railroad and automobile traffic, however, the state eventually needed a more permanent means of connecting its cities. Ferry service across Mount Hope Bay reportedly was under way by the early 18th century, and by the mid-19th

Conjectural view of the Mount Hope Bridge prepared shortly before construction.

century steam-powered vessels carried goods and people between Bristol, on the mainland, and Portsmouth, on the Island of Rhode Island, or, as it is also known, Aquidneck Island. This service continued into the 1920s, when increased automobile traffic prompted construction of a suspension bridge at the site. The Mount Hope Bridge Commission originally planned to build a steel cantilever bridge but, after the state legislature refused to fund the project, chose to build a less expensive suspension bridge design and to finance it by charging tolls. The design commission for the bridge went to David B. Steinman, acting through the firm of Robinson and Steinman, and field work began in December 1927. Completed in October 1929, the structure has a total length of 6,130 feet and a clear span between towers of 1,200 feet. During the Depression the privately held Mount Hope Bridge Company defaulted on its bonds and the bridge went through a succession of owners before finally coming under the control of the Rhode Island Turnpike and Bridge Authority in the mid-1960s. NR.

CRANSTON

The Arkwright Bridge is the largest 19th-century truss bridge that survives in Rhode Island. Located almost in the middle of the state, it is a single-span Pratt through truss with compression members formed by Phoenix columns. Phoenix columns were developed by Clark, Reeves and Company and popularized by its successor, the Phoenix Bridge Company, of Phoenixville, Pa. They are formed of wrought-iron members rolled along a circular arc. By riveting together these segments, it is possible to form a hollow column with a circular cross section. Structurally, this is an excellent shape for a compression member, and the Phoenix Bridge Company used it in many late 19th-century bridges and buildings. The only drawback was that the circular shape of the column was not well suited for making connections with other structural members, and the column was abandoned by the early 20th century.

The wrought-iron members of this bridge came from the Phoenix Bridge Company and were assembled on site by local contractors in accordance with plans prepared

■ **Arkwright Bridge**
Across the Pawtuxet River
On Hill Street
Dean and Westbrook
1888

Arkwright Bridge, displaying its distinctive nameplate.

by Dean and Westbrook, a New York engineering firm. The Arkwright Bridge is 128 feet long and has undergone little change or alteration in the past 100 years. It was built as part of a joint project by the towns of Cranston and Coventry, and the two municipalities still share in maintaining the bridge. NR.

HOPE

■ Gainer Memorial
(Scituate) Dam
Across the North Branch of
the Pawtuxet River
On State Route 12
Frank E. Winsor
1926

In 1915 the Rhode Island general assembly passed the Water Supply Act authorizing the city of Providence to build a large storage dam on the North Branch of the Pawtuxet River. Located about 10 miles southeast of the city, the elevation of the new dam allowed water to serve municipal users without the need for expensive pumping. In 1916 the city, acting through its Water Supply Board, began condemnation proceedings on approximately 15,000 acres of land that included more than six major cotton mills, 400 houses, seven schools and six churches within the "take" area for the Scituate Reservoir. Settlement on property and water diversion claims eventually cost the Water Supply Board more than $6 million. In retrospect, it was a small price to pay for a permanent relatively pure source of water. In addition, the city was able to build a small hydroelectric power plant as part of the new system.

Construction of the 3,200-foot-long, 180-foot-high earthfill dam continued for several years and was not completed until 1926. Built under the direction of Frank E. Winsor (1870–1939), an engineer who later served as president of the New England Water Works Association and as chief engineer for the Winsor Dam in central Massachusetts, the Scituate Reservoir has a storage capacity of more than 113,000 acre-feet. The structure's name refers to Joseph H. Gainer, mayor of Providence during planning and construction of the dam. Accessible to the general public, the Scituate Reservoir is the largest body of fresh water in Rhode Island. However, for sanitary reasons, no swimming, boating or other recreational use of the reservoir is allowed.

WICKFORD

Named in honor of Rhode Island's state engineer who served from 1912 until his death in 1925, the Hussey Bridge is a reinforced-concrete rainbow arch structure with an 80-foot span between abutments. Built to replace an all-metal Pratt through truss, the Hussey Bridge is distinguished by its use of steel hangar rods that carry the weight of the traffic deck to the arches. These steel rods are imbedded within wrought-iron pipes caulked with bitumen (a tarlike substance), a design developed to reduce the possibility that rust could destroy the hangars. The bridge is the only structure of this type in the state, and it has survived with little alteration. The picturesque span continues to carry highway traffic traveling along the southwest coast of Narragansett Bay.

■ **Clarence L. Hussey Memorial Bridge**
Across Wickford Cove
On State Route 1A
Rhode Island State Highway Commission
1925

Rainbow arch design of the Hussey Memorial Bridge.

■ ■ ■ ■ ■ ■ ■ VERMONT ■ ■ ■ ■ ■ ■ ■

EAST BERKSHIRE

The Jewett brothers — Sheldon and Savannah — of Montgomery, Vt., practiced the trade of covered-bridge building during much of the late 19th century, a time when metal truss technology was on the rise. According to the Vermont Division for Historic Preservation, the seven existing Jewett-built structures in Franklin County constitute the most extensive surviving record of any covered-bridge builders in the state. The Hopkins Bridge is a 91-foot-long, single-span Town lattice truss. It is the longest example of the Jewett brothers' work to survive. NR.

■ **Hopkins Covered Bridge**
Across the Trout River
Adjacent to State Route 118,
1 mile east of East Berkshire
Sheldon Jewett and Savannah Jewett, builders
1875

EAST CHARLOTTE

Blessed with extensive forests, Vermont fostered the development of numerous wooden covered-bridge builders during the 19th century. State residents are proud of this heritage, and they have worked hard to preserve many covered bridges on both aesthetic and historical grounds. This single-span, 86-foot-long Burr arch truss is one of three covered bridges preserved in Charlotte Township. Completed in 1849, it is among the oldest surviving Burr arch trusses in Vermont and it is

■ **Quinlan's Covered Bridge**
Across Lewis Creek
3 miles south of East Charlotte near Mt. Philo State Park
Leonard Sherman, builder
1849

Quinlan's Covered Bridge, a simple but functional design.

the only known covered bridge erected by local builder Leonard Sherman. The structure no longer carries highway traffic but it is accessible to pedestrians. NR.

EAST PUTNEY

■ **East Putney Brook Stone Arch Bridge**
Across East Putney Brook
Adjacent to River Road
(Town Route 7)
James Otis Follett, builder
1902

Stretching 30 feet across East Putney Brook, this stone arch structure is a small reminder that Vermont's use of masonry for highway bridges continued long after wood, iron and steel had proven their practicality. It was built by James Otis Follett, a farmer and self-trained mason who was born in East Jamaica and lived in nearby Townshend. As a local road commissioner for many years, Follett became involved in building and maintaining highways in the region, subsequently specializing in small stone arch highway bridges. The East Putney span is not a landmark in state-of-the-art engineering design; rather, it is a monument to an older construction tradition in which designs were based on experience and intuition instead of mathematical analysis. The bridge served local highway traffic until the mid-1960s when it was bypassed by a new highway. It now carries only pedestrian traffic. NR.

East Putney Brook Bridge, one of James Otis Follett's best stone arch structures.

HIGHGATE FALLS

Located about five miles south of the United States–Canada border, the Highgate Falls Bridge consists of a 215-foot-span, lenticular through truss and an 80-foot-span, lenticular pony truss. Among the largest surviving lenticular truss bridges in the United States, it is a significant reminder that, despite the plethora of covered bridges, iron spans played a significant role in the development of Vermont's agriculture and industrial economy. NR.

■ **Highgate Falls Bridge**
Across the Missisquoi River
On State Route 2
Berlin Iron Bridge Company
1887

MIDDLEBURY

This remarkable structure is one of the few surviving double-barrel covered bridges in the United States. Instead of having a roadway that passes between a single pair of trusses, the bridge has a third truss that extends down the middle of the structure. This provides the bridge with two distinct roadways that can carry two-way traffic without fear of collision. The original design of the Pulp Mill Bridge, one of the oldest to survive in Vermont, consisted of a 181-foot-long, single-span Burr arch truss, but the structure has subsequently been strengthened by the addition of two piers in the middle of the river. It is still used for light vehicular traffic. NR.

■ **Pulp Mill**
Covered Bridge
Across Otter Creek
½ mile north of State
Route 23
c. 1830

Left: Double-barrel design of the Pulp Mill Covered Bridge. Below: Flint Covered Bridge.

NORTH TUNBRIDGE

The Flint Covered Bridge is the oldest of five covered bridges that span the First Branch of the White River within a distance of about seven miles. The single-span, 88-foot-long structure is a good example of a large queen-post truss; a variant of the king-post truss, this design uses two, rather than one, vertical members. It was renovated in 1969 and still carries a small amount of local traffic. NR.

■ **Flint Covered Bridge**
Across the First Branch of the White River
Adjacent to State Route 110, 3 miles north of North Tunbridge
1845

PITTSFORD

■ **Hammond Covered Bridge**
Across Otter Creek
1 mile west of U.S. Route 7,
near Florence Station
Asa Nourse, builder
1842

Left: Hammond Covered
Bridge. Right: Follett's
Sacketts Brook Bridge.

Today, this 139-foot-long, single-span Town lattice truss is preserved by the state of Vermont as a historic site, but the structure almost suffered a cruel fate when the great flood of November 1927 knocked it off its abutments and sent it floating downstream. That flood destroyed scores of covered bridges in Vermont, but the Hammond Bridge was one of the lucky ones. Before the waters had substantially receded, city officials were able to tow the bridge back to its site and eventually return it to service. Presumably the stiffness of the Town lattice design helped prevent the bridge from warping or buckling throughout the ordeal. NR.

PUTNEY

■ **Sacketts Brook Stone Arch Bridge**
Across Sacketts Brook
On Town Highway 25
James Otis Follett, builder
1906

Completed in 1906, the Sacketts Brook Bridge is a 29-foot-span, dry-laid masonry structure. Of the more than 30 stone arch bridges built between 1890 and 1911 by local builder James Otis Follett, this is one of only a few that still carries highway traffic. Its simple, yet effective, design stands as testimony to the inherent strength of masonry in compression and to the suitability of stone arches for bridge construction. NR.

WINDSOR

■ **Ascutney Dam**
Across Mill Brook
South of State Route 131
Simeon Cobb and Joseph Mason, builders
1834

The flourishing of 19th-century mill construction in New England depended largely on the development of water power along the rivers and streams in the region. Dams were required to divert water out of riverbeds so that it could be channeled to waterwheels or turbines. In certain circumstances, dams also were built to store water flowing in a stream so that it could be released at times when downstream mill owners could make maximum use of it.

Among the earliest permanent storage dams in America is the Ascutney Dam in Windsor, only a few miles from the Connecticut River. As documented by Edwin Battison, director of the nearby American Precision Museum, the 40-foot-high, 250-foot-long curved gravity structure was built by the Ascutney Mill Dam Company to increase the productivity and, thus, the value of water power sites on lower Mill Brook. Construction began in April 1834 with the removal of a small dam previously built at the site. The first granite blocks, obtained from

Ascutney Dam, with the concrete coping that helps prevent masonry erosion when the reservoir overtops during floods.

nearby quarries, were laid in June, and by November the structure was complete. Although the Ascutney Mill Dam Company experienced problems following the financial panic of 1837, it continued to operate the dam throughout the 19th century. In the 20th century the structure served as part of a hydroelectric power generating system, and a concrete coping, or covering, was placed on the top of the structure to help ensure its stability in times of flood. Aside from this, the dam retains much of its original structural integrity and stands as one of the oldest dams in the United States. ASCE.

WOODSTOCK

Since 1797 a bridge has crossed the Ottauquechee River in Woodstock. During the 19th century at least six wooden bridges at the site, many built by the Royalton and Woodstock Turnpike Company, either floated away in floods or collapsed from decay. In 1869 a town meeting approved construction of a new, longer iron bridge at a higher elevation above the stream. The contract for the new 110-foot-long span went to the Boston-based National Bridge and Iron Company, and by early 1870 it was open for traffic.

The wrought-iron Parker truss had functioned successfully for more than 100 years when it became the focus of a major controversy related to the federal funding of highway bridge replacement projects (see page 65). The citizens of Woodstock wanted to keep the 18-foot-wide bridge in operation, but the state believed that the narrow structure constituted a safety hazard, especially because it was part of a federal-aid system highway. In addition, federal bridge replacement standards complicated the problem by requiring that federally funded projects result in a rehabilitated bridge at least 30 feet wide. The situation attracted considerable publicity, but, more important, it brought together preservationists and highway officials to discuss the problems of reconciling modern transportation needs with the goals of historic bridge preservation. In the end, a compromise was reached that provided federal funding to widen the bridge to 24 feet along a new alignment across the river and strengthen the roadway with concrete-girder supports. Historic bridge aficionados might grumble about these alterations, but the wrought-iron Parker truss is still in place, looking similar to its original 19th-century appearance. NR district.

■ **Elm Street Bridge**
Across the
Ottauquechee River
On State Route 12
National Bridge and Iron
Company
1870, 1979

End post of the Elm Street Bridge with National Bridge and Iron Works nameplate.

MID-ATLANTIC

Night view of the Brooklyn Bridge across the East River in New York City. John Roebling's masterpiece is probably one of the most famous, and beloved, bridges in the world.

■ ■ ■ ■ ■ ■ ■ DELAWARE ■ ■ ■ ■ ◢ ■ ■

NEW CASTLE

■ **Delaware Memorial Bridge**
Across the Delaware River
On Interstate 95
Howard, Needles, Tammen and Bergendoff; Othmar H. Ammann
1951, 1968

Today, it is almost impossible to imagine what travel up and down the Mid-Atlantic coast would be like without the Delaware Memorial Bridge. But before 1951, motorists on U.S. Route 40 had to use a ferry to cross the Delaware River between New Castle, Del., and Pennsville. N.J. — often a frustrating source of delay (as the author's mother can attest). This situation changed completely with the opening of the Delaware Memorial Bridge, intended to facilitate transport of the huge amount of traffic along the eastern seaboard. The 1951 bridge is a steel suspension structure with a clear span of 2,150 feet between the towers. It uses a riveted Warren truss for the traffic deck to help resist aerodynamic forces. Swiss-trained engineer Othmar H. Ammann (1879–1967) designed several other major American bridges in the region, including the George Washington Bridge (1931) in New York City, and he undertook the commission for the Delaware Memorial Bridge in concert with the large engineering firm of Howard, Needles, Tammen and Bergendoff. In response to the increase of highway traffic in the 1960s, a new bridge was built directly upstream from the original span and thus doubled the traffic capacity. Completed in 1968, the new structure used the same plans as the 1951 bridge, and, without close physical examination, it is impossible to distinguish visually between the two spans. Today, each bridge continues in heavy use carrying one-way traffic.

Original Delaware Memorial Bridge before construction of the parallel span in the 1960s.

WILMINGTON

■ **Pennsylvania Railroad Brick Arch Viaduct**
Along the Amtrak right-of-way
South of the Wilmington Amtrak Station at Liberty Street
William H. Brown
1908

In the late 19th century America's largest railroads were flush with revenue and in a position to take on large projects to improve the quality of their track system. This work would increase their carrying capacity and, they hoped, reduce future maintenance costs. In particular, railroads were interested in eliminating highway grade crossings that were the source of serious accidents. In 1901 the Pennsylvania Railroad, acting through its subsidiary, the Philadelphia, Baltimore and Potomac

Railroad, began a project to build an elevated viaduct through the city of Wilmington to remove grade crossings and thus provide speedier, more reliable service. A key part of this plan involved building a mile-long arch viaduct on the soft ground at the southern edge of the city. Originally, the arches were to have been built of stone, but material shortages dictated the use of brick. Each of the more than 70 arches has a clear span of 41 feet. The structure still carries traffic along Amtrak's Northeast Corridor.

Top and center: Wilmington's viaduct under construction, c. 1907. Above: Contemporary view of the completed structure.

WOODDALE

Covered-bridge historian Richard S. Allen estimates that at one time there were at least 30 covered bridges in Delaware. Since the early 20th century, bridge replacement projects, floods and arson have taken such a toll that today only two covered bridges survive in the state. The longest of these is the 72-foot-long Town lattice truss across Red Clay Creek in Wooddale. NR.

■ **Wooddale Bridge**
Across Red Clay Creek
½ mile north of State
Route 48
c. 1860

■ ■ ■ ■   DISTRICT OF COLUMBIA   ■ ■ ■ ■

■ Arlington Memorial Bridge

Across the Potomac River
On U.S. Route 50 between
the Lincoln Memorial and
Arlington National Cemetery
McKim, Mead and White,
with John L. Nagle
1932

Located between two of the capital's major tourist attractions, Arlington Memorial Bridge was started 40 years after a bridge at this site was first officially studied and even longer after Andrew Jackson first suggested such a span to symbolize the union between the North and South. Various proposals to cross the Potomac River here were developed and abandoned over the years, including a twin-towered design chosen through an architectural competition in 1900. Eventually, with the creation of the McMillan Park Commission and U.S. Commission of Fine Arts to expand the ceremonial core of Washington, plans turned to creating a bridge that

Above: Stately, granite-clad Arlington Memorial Bridge across the Potomac. Right: Equestrian sculptures adorning the entrance near the Lincoln Memorial. The sculptures were cast in Italy as a gift to the American people.

would be integrated into the monumental new plan for the capital. Providing a dramatic entrance to Arlington National Cemetery, the Memorial Bridge is particularly notable for its role in extending the axial plan of the Mall across the river.

The bridge, begun in 1926, owes its neoclassical style to the noted architects McKim, Mead and White (Stanford White served on the 1900 design competition jury). The ornamented masonry facing lends dignity to the bridge and relates it to the Lincoln Memorial (1922, Henry Bacon). Many people consider this the most beautiful bridge in Washington, while others have decried the facing as an ill-conceived attempt to obscure the purity of the structural design. Elizabeth Mock, for example, described it in *The Architecture of Bridges* (Museum of Modern Art, 1949) as "designed in Washington's usual pompous neo-classic manner." With nine arches, the reinforced-concrete span is 2,138 feet long and 60 feet wide with 15-foot walkways. The center arch is a double-leaf bascule bridge built to provide clearance for ships traveling to the old port of Georgetown, about a mile upstream. Although built of steel, it was designed to blend in with its neighboring masonry-clad arches. At 216 feet long, the draw span is among the longest in the world but has not been opened for several years. Completing the bridge's ceremonial character are sculpted eagles and bison by Paul C. Jennewein as well as golden equestrian figures by Leo Friedlander. The bridge remains in heavy use carrying U.S. Route 50 across the Potomac. NR.

Now the oldest surviving bridge across the Potomac in Washington, the Key Bridge was named for Francis Scott Key, lyricist of the "Star Spangled Banner." In the early 19th century Key lived near the Georgetown end of the bridge in a house that later was razed for an expressway ramp. Built by the U.S. Army Corps of Engineers from 1917 to 1923, the bridge is a seven-span, three-ribbed, open-spandrel arch design constructed of reinforced concrete. More than 1,700 feet long, the bridge connects the historic Georgetown area with Virginia to the south. Its crossing is located a short distance downstream from the site of the 1830s Potomac Aqueduct, built to connect

■ Key Bridge
Across the Potomac River
Between Georgetown and
Rosslyn, Va.
Nathan C. Wyeth
1923

View of the Key Bridge's upstream facade from the Georgetown shoreline.

Above: Key Mansion (demolished), once located near the site of the Key Bridge. Right: Wooden centering for one of the bridge's arches.

the Chesapeake and Ohio Canal with Alexandria. Used as a military bridge during the Civil War, the aqueduct was replaced in the 1880s by a metal truss bridge that used the same stone piers built for the 1830 structure (one of which is still visible near the Virginia shore). The Key Bridge took the place of the metal truss bridge, but its piers were built on completely new foundations. The District of Columbia transportation department recently refurbished the deck of the bridge and modernized the walkways and railings. It is now ready to carry traffic until well into the 21st century.

■ **Taft Bridge**
Across Rock Creek Park
On Connecticut Avenue near Calvert Street
George S. Morison and Edward Casey
1907

Rock Creek is a small tributary of the Potomac River that cuts through northwest Washington, D.C. For much of its course, the stream flows through a relatively deep valley that impeded transportation over the creek during the 19th century. In the early 20th century real estate interests pushed for construction of a high-level bridge across the creek that would open up new areas, including Chevy Chase, Md., for residential development. Construction of the high-level Taft Bridge began in 1897, and it was opened for traffic in 1907. A five-span, open-spandrel concrete arch bridge with a total length of more than 900 feet, it was named in 1931 for President William Howard Taft. Except for a small amount of steel in the deck, the structure consists entirely of unreinforced concrete. Because of the excessive amount of material in the design, it was dubbed the "Million Dollar Bridge" for

Aerial view of Rock Creek Park, with the Taft Bridge in the foreground, c. 1930.

its high cost. Designed by George S. Morison and constructed under the supervision of architect Edward Casey, the Taft Bridge is Morison's only major concrete structure. A pair of lions sculpted by R. Hinton Perry guards each end of the bridge. The structure is scheduled for renovation in the near future, but this should have little impact on its overall visual appearance.

High-level Taft Bridge on Connecticut Avenue shortly after construction, c. 1910.

■ ■ ■ ■ ■   MARYLAND   ■ ■ ■ ■ ■

BALTIMORE

Built for the Baltimore and Ohio Railroad, America's first railroad, the Carrollton Viaduct is the oldest railroad bridge in the country. Its single-span, 100-foot-long hemispherical stone arch is reminiscent of Roman arch bridge design, and, as with several Roman structures, it appears capable of standing for hundreds more years. Named after Charles Carroll, the last living signer of the Declaration of Independence, the Carrollton Viaduct still carries traffic for the B&O Railroad — testimony to the inherent strength of masonry arch designs and the great care taken in the late 1820s to ensure its long-term structural integrity. Although in photographs the bridge appears to lie in a bucolic rural setting, it is actually in the midst of a commercially developed part of Baltimore. NR, ASCE.

■ **Carrollton Viaduct**
Across Gwynns Falls
On the Baltimore and
Ohio Railroad line, ½
mile south of U.S. Route 1
Caspar Weaver and
James Lloyd
1829

Carrollton Viaduct in a seemingly rural setting.

■ **Druid Lake Dam**
In Druid Hill Park
West of Interstate 83
Robert K. Martin
1870

In 1864 the Baltimore Water Works began construction of a large storage dam to serve the needs of the rapidly growing city. Built over a period of six years, the Druid Lake Dam is an earthen structure with a maximum height of 119 feet and a length of more than 800 feet. The dam appears to be little more than a natural hill, not a man-made structure, but it is a massive work with a maximum base width of 640 feet. At the time of completion it was the largest earthen dam in the United States. It remains in use as an important component of Baltimore's water supply system. NR district. ASCE.

BETHESDA

■ **Cabin John Aqueduct**
Across Cabin John Creek
On MacArthur Boulevard
Montgomery C. Meigs
1864

During the early 19th century Washington, D.C., was a small city on the Potomac River without a large population. Although it served as the nation's capital, it did not initially require a major public works system as water was drawn from local springs and wells. This supply proved adequate during the city's early years, but by the mid-19th century a new source of uncontaminated water was required. Beginning in 1857 the U.S. Army Corps of Engineers began work on a 15-mile-long aqueduct to bring water to the city from Great Falls on the Potomac River. The aqueduct system was developed by the innovative engineer Montgomery C. Meigs (1816–92), who supervised construction of many Washington struc-

Right: Construction of the Cabin John Aqueduct. Below: Aqueduct in its original setting, c. 1910. Today, a four-lane highway passes under the arch.

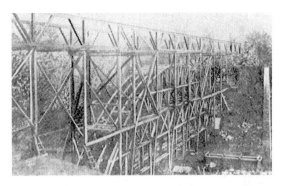

tures including the old Pension Building (1885), now the home of the National Building Museum.

The only major natural obstacle crossed by the aqueduct is Cabin John Creek, a small stream flowing through a deep valley about six miles from the falls. Meigs crossed this valley with a bold 220-foot-long, single-span masonry arch bridge built with sandstone from the nearby Seneca quarries. The random ashlar facing contrasts with a simple band of dressed stone trim along the arch. Despite delays caused by the Civil War, the structure was completed in 1864. Today, it still carries a portion of Washington's water supply as well as one lane of light vehicular traffic. It survives as America's longest single-span masonry arch bridge. NR, ASCE.

CONOWINGO

The Susquehanna River drains most of central Pennsylvania, as well as a considerable portion of south-central New York, and is the major source of fresh water for the Chesapeake Bay. Located only about 10 miles upstream from where the river joins the bay, the Conowingo Dam was built to develop the huge hydraulic power potential inherent in the lower reaches of the river. The Philadelphia Electric Company began planning a major hydroelectric power plant at Conowingo in the early 1920s to deliver power to eastern Pennsylvania and parts of New Jersey. Acting through the Susquehanna Power Company and the Philadelphia Electric Power Company, both wholly owned corporate subsidiaries set up to help eliminate problems associated with the multistate character of the project, the Philadelphia-based power company started work on the dam in early 1926.

The design commission for the Conowingo Dam was given to the Boston-based firm of Stone and Webster. Formed in the early 1890s by Charles Stone and Edwin Webster, two graduates of the Massachusetts Insititute of Technology, the firm specialized in designing, building, financing and operating electric power systems through-

■ **Conowingo Dam**
Across the Susquehanna River
On U.S. Route 1
Stone and Webster
1928

Conowingo Dam and powerhouse, as depicted by painter Daniel Garber, c. 1930.

Conowingo Dam showing the full reservoir and U.S. Route 1 running across the top of the structure.

out the United States. The company retained the confidence of wealthy capitalists in Boston and New York who provided the economic leverage to develop large regional power systems. Financed entirely by private capital, the Conowingo Dam is 4,700 feet long with a maximum height of 105 feet. It is a straight-crested, concrete gravity structure with an overflow spillway more than 2,700 feet long. At the time of its completion in 1928, the dam had an installed generating capacity of 378,000 horsepower, an amount second only to the hydroelectric facilities at Niagara Falls. The dam is still in active use and carries U.S. Route 1 over the Susquehanna River.

ELKRIDGE

■ **Thomas Viaduct**
Across the Patapsco River
On the Baltimore and Ohio
Railroad line, near U.S.
Route 1
Benjamin H. Latrobe II
1835

Multispan Thomas Viaduct south of Baltimore, as depicted in a 19th-century print.

Considered one of the oldest operating railroad bridges in the United States, the Thomas Viaduct is an eight-span masonry arch bridge on the Baltimore and Ohio Railroad line between Baltimore and Washington, D.C. Built with a slight four-degree curve to accommodate the railroad right-of-way to the local topography, the viaduct has a total length of 612 feet and extends 62 feet above the creekbed. Designed by the son of Benjamin H. Latrobe, one of America's most famous architect-engineers of the early 19th century, the bridge is named after Philip Thomas, the first president of the B&O Railroad. The Thomas Viaduct is an enduring accomplishment as it is still in daily service carrying freight and passenger traffic after more than 150 years. NR.

GRANTSVILLE

In 1806 Congress authorized construction of the National Road from the Potomac River watershed to the Ohio River and points west, and by 1811 construction began west of Cumberland, Md. One of the first major bridges of this huge public works endeavor was the Casselman's Bridge, a 354-foot-long, single-span masonry arch bridge with clearance between abutments of approximately 100 feet. The center of the span rises more than 30 feet above the usual stream level and thus required rather steep approaches. The bridge is an extremely sturdy, well-made structure that necessitated a considerable amount of skilled labor to erect. For more than 100 years it carried wagon and automobile traffic, but today it is preserved within the confines of a county park and is accessible only by foot. NR.

■ **Casselman's Bridge**
Across the Casselman River
Adjacent to U.S. Route 40
1813

Left: Casselman's Bridge showing a slight peak at the top of the arch. Right: Stone arch span when it still carried highway traffic, c. 1910.

MIDDLETOWN

This structure is an excellent example of a wrought-iron, pin-connected, double-intersection Pratt through truss. It is a single-span bridge with a length of approximately 120 feet. Aside from the addition of steel guard rails, the bridge is practically unaltered from its original construction. It still remains in use carrying a small amount of light vehicular traffic. NR.

■ **Poffenberger Road Bridge**
Across Catoctin Creek
On Poffenberger Road,
1 mile south of State
Route 17
Penn Bridge Company
1878

Poffenberger Road Bridge, with its simple portal decoration.

Above: Old Mill Road Bridge, with its masonry abutments. Right: Cast-iron plate with names of 1882 county commissioners. Note that "Pittsburg" is spelled without an "h."

ROCKY RIDGE

■ Old Mill Road Bridge
Across Owens Creek
On Old Mill Road,
½ mile south of State Route 77
Pittsburg Bridge Company
1882

The Old Mill Road Bridge is a 69-foot-long, single-span, pin-connected Pratt through truss built to provide access to a small rural grist mill. The mill has long since disappeared, but the bridge survives and is used to carry a small amount of highway traffic. Although it is a relatively small structure, the cast-iron nameplate lists the names of all the county commissioners in office at the time the contract was let, reminding voters of the public officials responsible for the bridge's construction. Apparently the bridge company provided this political advertising in hopes that the commissioners would remember the favor the next time they planned to build a new truss bridge. In the late 19th and early 20th centuries, nameplates were commonly applied to bridges and often listed public officials. NR.

SANDY POINT

Throughout the first two centuries of Maryland's history, the Chesapeake Bay served as a vital trade and transportation route. The huge bay, which almost splits the state of Maryland in two, prompted the growth of extensive maritime industries in the region. But, as communities became more dependent on railroads and automobiles, the Chesapeake Bay was seen as a barrier between Maryland's Eastern Shore region and the rest of the state on the west side of the bay. Ferry service helped link the two sides of the bay, but many considered this an inadequate long-term solution to the problem. Following an aborted attempt to finance private construction of a bridge in the late 1920s (this plan collapsed because of the Depression), the state legislature by the end of the 1930s authorized detailed planning for a long-span structure between Sandy Point and Kent Island. Construction was delayed by World War II, but in 1947 the legislature gave final approval for the project and construction commenced in January 1949. Opened for traffic in July 1952, the bridge has a total length (including approach spans) of four miles. The suspension span has a distance of 1,600 feet between towers and provides a clearance of 186.5 feet above mean high water.

The bridge proved to be a great success and by the 1960s had prompted a huge surge in resort development along the Atlantic coast of the Delmarva Peninsula. This led to the construction of a second bridge built only 450 feet south of the original crossing and completed in 1973. This parallel span is similar in design to the original structure and includes a 1,600-foot suspension bridge with a vertical clearance of 186.5 feet. However, motorists can visually distinguish it from the earlier structure because of the shape of the bracing for the towers. The original bridge uses X-shaped cross bracing, while the 1973 design uses simple horizontal girders for tower bracing. In addition, the traffic deck for the original span is a through truss, while the later bridge carries traffic on a deck truss. Today, the bridge carries heavy amounts of traffic (more than 12 million vehicles annually) and is well known to residents of the Baltimore-Washington area who trek to the ocean on hot summer weekends. The Chesapeake Bay Bridge is officially named in honor of William Preston Lane, Jr., a former governor of Maryland.

■ **William Preston Lane, Jr., Memorial (Chesapeake Bay) Bridge**
Across the Chesapeake Bay
On U.S. Route 50
J. E. Greiner Company
1952, 1973

Lane Memorial Bridge across the Chesapeake Bay.

SAVAGE

The growth of America's railroads spurred the development of new types of bridges to carry the "iron horse" over innumerable rivers and streams. Wood and masonry spans were first used by railroad companies, but by the early 1850s engineers were constructing iron bridges on several main lines. As documented by Robert Vogel, curator of mechanical and civil engineering at the Smithsonian Institution, the Bollman suspension truss became the first all-metal truss bridge used consistently

■ **Bollman Truss Bridge**
Across the Little Patuxent River
1 mile west of U.S. Route 1, adjacent to Savage Mills
Wendel Bollman
1869

One span of the Bollman Truss Bridge, with Savage Mills in the background.

by an American railroad. Developed by Wendel Bollman for use on the Baltimore and Ohio Railroad, the Bollman truss consists of cast-iron compression members and wrought-iron tension members. Each panel point on the lower chord is supported by two diagonal tension members connected to the top of both end posts. In this way each floor beam of the truss is suspended from the cast-iron end posts at each abutment. The major structural advantage of the design is that a failure of a diagonal tension member will cause the collapse of only a single floor beam; with most other truss designs, failure of a diagonal will cause the entire span to collapse.

Bollman built the first of his trusses for the B&O at Savage in 1850. The structure now at the site, however, was fabricated in 1869 for use on the railroad's main line. The company moved the bridge to Savage in the late 19th century for use on a small spur line serving Savage Mills. It is a two-span bridge with a total length of 160 feet; each individual span is 80 feet long. Today, the bridge at Savage is the only known surviving Bollman truss in the world. It is maintained by Howard County as part of a regional park system and is readily accessible to visitors. NR, ASCE.

■ ■ ■ ■ ■ ■   NEW JERSEY   ■ ■ ■ ■ ■ ■

BAYONNE

■ **Bayonne Arch Bridge**
Across Kill van Kull
On State Route 440
Othmar H. Ammann
and Allston Dana
1931

In the 1920s the growth of automobile traffic in the metropolitan New York City area prompted serious interest in building highway bridges to connect Staten Island, N.Y., with New Jersey. Acting through the bistate Port Authority of New York and New Jersey, in 1927 both state legislatures authorized construction of a long-span bridge across the Kill van Kull along the island's north shore. Because of navigation requirements, the structure had to have a clear span between piers of more than 1,600 feet, thus dictating either a steel suspension or a long-span through arch design. In the end the determining factor proved to be the dense, fine-grained bedrock on both shores of the site. This bedrock provided an excellent foundation for supporting a massive arch structure and, just as important, would have seriously impeded the extensive excavation required for a suspension bridge's anchorages.

These geological considerations subsequently led to the construction of a 1,675-foot-long steel arch span. Originally, the Port Authority planned to build masonry-sheathed towers at the abutments, similar to those of the Hell Gate Bridge (1917) in New York City. These plans were later abandoned, however, apparently for financial reasons. The structure remains a major transportation route between New Jersey and Staten Island. ASCE.

CLINTON

Designed by Francis C. Lowthorp and built by William Cowin of Lambertville, N.J., the Main Street bridge across the Raritan River's South Branch is a distinctive cast-iron and wrought-iron structure. Lowthorp was a firm advocate of cast iron as a structural material, and in the 1850s and 1860s he patented several methods of using it in bridges. The basic configuration of Lowthorp's bridge in Clinton is a Pratt truss that uses cast-iron compression members and wrought-iron tension members, but the form of the connections distinguishes it as a Lowthorp design. It consists of two 85-foot-long pony trusses. Some of the original vertical members have been replaced with steel I beams, but, in general, the structure retains its historical integrity. Visually, its most striking feature is the design of the cast-iron, Italianate end posts. The bridge is still used for light vehicular traffic and forms an integral part of Clinton's downtown historic district. NR district.

■ **Lowthorp Truss Bridge**
Across the South Branch of the Raritan River
On Main Street
Francis C. Lowthorp
1870

Left: Lowthorp Truss Bridge in Clinton, N.J. Right: Detail of the cast-iron Italianate end post.

JERSEY CITY

■ **Pulaski Skyway**
Across the Passaic and
Hackensack Rivers
On U.S. Route 1
Sigvald Johannesson
1932

With the growth of automobile traffic around New York City during the 1920s, improvements were needed in the highway system that fed into the newly completed Lincoln Tunnel under the Hudson River. To provide a high-level crossing of the marshy Hackensack Meadowlands between Jersey City and Newark, construction began in 1930 on the Pulaski Skyway. This 16,000-foot-long viaduct consists of numerous short-span Pratt deck trusses and two 1,250-foot-long steel cantilever through trusses that span the Passaic and Hackensack rivers. The Pulaski Skyway initially carried traffic along the easternmost section of the transcontinental Lincoln Highway. Since then the Lincoln Highway (U.S. Route 30) has changed routes—it now terminates, or begins, in Atlantic City — and the viaduct has become part of U.S. Route 1.

PRINCETON

■ **Stony Brook
Arch Bridge**
Across Stony Brook
On. U.S. Route 206
c. 1795

Located about one mile south of Princeton's town center, the Stony Brook Bridge is a three-span, masonry arch structure with a total length of about 150 feet. Built in the late 18th century, the bridge was part of a main highway route between Princeton and Trenton. The northwest

Three-span masonry arch bridge still in service south of Princeton.

abutment wall, on the Princeton Battlefield, is imbedded in remnants of the early 18th-century Worth's Grist Mill. The bridge remains in use carrying traffic along a major regional highway. NR.

TRENTON

The Calhoun Street Bridge is a seven-span, pin-connected Pratt through truss with a total length of 1,280 feet. Its compression members consist of the distinctive circular Phoenix columns developed by Clark, Reeves and Company and popularized by its successor, the Phoenix Bridge Company. In 1885 the company considered the Trenton City Bridge, as it was sometimes called, significant enough to warrant a collotype illustration in a major promotional catalog. This catalog described it as representing "the most advanced system of construction as applied to the ordinary roadway bridge." The structure later became part of the original route of the Lincoln Highway when it cut across New Jersey to New York City. Today, the Calhoun Street Bridge is owned by the Delaware River Joint Toll Commission and is open to light vehicular traffic. Plans have been made to remove it from service in the future, but it will almost certainly be preserved in place as a pedestrian-bicycle bridge.

■ **Calhoun Street Bridge**
Across the Delaware River
On Calhoun Street
Phoenix Bridge Company
1885

Above: Contemporary view of the Calhoun Street Bridge. Left: Calhoun Street Bridge when it was part of the Lincoln Highway, c. 1925.

BUFFALO

■ Peace Bridge
Across the Niagara River
Adjacent to Interstate 190
Edward P. Lupfer
1927

The Peace Bridge is a good example of an early 20th-century, multiple-arch, steel-plate girder bridge. Built to provide an international crossing between Buffalo and Fort Erie in the province of Ontario, Canada, the structure's deck arches vary in length from 346 to 423 feet. There is also a riveted Pennsylvania through truss adjacent to the American shore that provides clearance for ship navigation. The bridge's distinctive name refers to the long-time peaceful relations between the United States and Canada. It is the most traveled crossing between the two countries.

CLAVERACK

■ Shaw Bridge
Across Claverack Creek
On Van Wyck Lane, near
State Route 9H
John D. Hutchinson
1870

Squire Whipple (1804–88) is remembered by engineering historians for his landmark 1847 treatise on truss bridge analysis and for his development of large double-intersection Pratt trusses (often simply called Whipple trusses). In 1841 he also patented a cast-iron and wrought-iron bowstring arch truss design that achieved great renown in the mid-19th century. This bowstring design is considered the first successful all-iron truss bridge design in the U.S., and it represents a major part of Whipple's engineering legacy.

Often used for bridges crossing the Erie Canal, the bowstring design found widespread use in New York State and elsewhere. The Shaw Bridge, in the central Hudson River Valley, is the only two-span Whipple bowstring arch truss that survives in the United States. As documented by William P. Chamberlin, an expert on New York's bridge heritage, the structure was built by John D. Hutchinson of Troy, N.Y., and located along the route of the original post road from New York to Albany. The Hutchinson family, including John D. and his father, began building Whipple bowstring designs in the 1850s. Interestingly, although the 160-foot-long Shaw Bridge was erected almost 30 years after Whipple's patent application, it follows the patent design almost to the letter. Perhaps more than anything else, this bridge is the best testimony to the quality of the design. The Shaw Bridge is named for William Shaw, a prominent resident of Claverack who lived near the bridge in the late 19th century. The structure continues to carry local highway traffic.

Detail of the bottom-chord connection used on the Shaw Bridge.

CROTON-ON-HUDSON

■ New Croton Dam
Across the Croton River
Near State Route 29
Alphonse Fteley
1907

In the 1830s New York City began building one of America's largest and most ambitious water supply systems. Under the guidance of its chief engineer, John B. Jervis, the city constructed an earthen dam on the Croton River about 40 miles north of Manhattan. The Croton Reservoir was then connected to the city by a large aqueduct designed to serve municipal water needs for

Above: Peace Bridge, with its steel-plate girder arches. Left: Cast-iron arch design developed by Squire Whipple and used for the Shaw Bridge. Below: Aerial view of the New Croton Dam, with its spillway on the far left.

Detail of the masonry work
on the New Croton Dam.

decades to come. Completed in 1842, the Croton Dam
and Aqueduct provided exemplary service for many
years. But by the 1880s the rapidly growing city required
more water than Jervis's system could supply. Along with
a second aqueduct pipeline, the city began planning for a
much larger dam on the Croton River to store floodwaters
that previously flowed into the Hudson River. Using
recent French designs for large-scale masonry gravity
dams as a model, the aqueduct commission developed
plans for a 262-foot-high structure at a location known as
the Quaker Bridge site. In the 1890s the commission
relocated the proposed dam to a new site approximately
three miles below the original Croton Dam.

Using the Quaker Bridge plans for its basic design, the
New Croton Dam is a straight-crested masonry gravity
dam with a maximum height of 297 feet and a total
length, including the spillway, of more than 2,000 feet.
Work on the massive structure began in 1892 and
continued for almost 15 years. The structure's design by
Alphonse Fteley (1837–1903) became a standard refer-
ence for gravity dams, and the "Croton Profile" achieved
international recognition. At the time of its completion,
the New Croton Dam was the tallest dam in the world.
Today, it remains in service as a critical component in
New York City's water supply system. ASCE.

FORT HUNTER

■ **Schoharie Creek Aqueduct**
Across Schoharie Creek
Part of Schoharie Crossing
State Historic Site
John B. Jervis
1841

In 1815 the completion of the original Erie Canal across
upstate New York helped open up large sections of the
western United States to settlement and commercial
development. By the late 1830s the state of New York
wished to expand the canal's capacity and began a major
program to upgrade the system. One of the most
important parts of this improvement program involved
construction of a canal aqueduct across Schoharie

Creek. Previously, canal boats actually entered the river and were pulled across the treacherous stream by a system of ropes and windlasses.

In 1839 work began on a wood and masonry aqueduct that would carry the canal over the creek and provide much quicker and safer passage. Completed two years later, this structure included a wooden trough for the canal proper and a stone arch bridge designed to support the adjacent towpath. After the opening of the New York State Barge Canal in the early 20th century, the Erie Canal went out of service, and five arches from the towpath bridge were subsequently demolished. In addition, the wooden canal trough was also removed. Today, approximately 415 feet of the arch bridge, encompassing nine arches with individual spans of 45 feet, survive and are preserved as part of Fort Hunter State Park.

Surviving masonry arches of the Schoharie Creek Aqueduct.

NEW YORK CITY

In the 1930s New York City began an extensive program to upgrade and expand its highway system. Much of this work was undertaken by Robert Moses's Triborough Bridge and Tunnel Authority, an autonomous state agency authorized to sell bonds, build bridges and collect tolls. After constructing the Triborough Bridge in 1936, the authority continued its growth in the late 1930s by building the 2,300-foot-long Bronx-Whitestone Bridge across the eastern end of the East River. The bridge's original design included an extremely thin steel girder deck. In 1940 the newly built Tacoma Narrows Bridge (Leon Moisseiff, engineer) in Tacoma, Wash., collapsed in a wind storm, causing consternation among suspension bridge engineers throughout the world. The Tacoma failure was related to aerodynamic forces acting on the structure's thin traffic deck, which set up the oscillations that destroyed the span. Once Othmar Ammann realized

■ **Bronx-Whitestone Bridge**
Across the East River
On Interstate 678
Othmar H. Ammann and
Allston Dana
1939

the danger posed by the thin steel-girder traffic deck, the Bridge and Tunnel Authority strengthened the Bronx-Whitestone Bridge by superimposing a new steel truss onto the deck. In addition, cable stays were added between the towers and the deck to prevent twisting. Today, the rehabilitated bridge remains a vital component in the city's highway system.

■ **Brooklyn Bridge**
Across the East River
John A. Roebling and
Washington Roebling
1883, 1952

With the celebration of its 100th anniversary in 1983, the Brooklyn Bridge was the object of more affection and attention than any bridge in history. But even after all the excitement has subsided, the structure still strikes even the most casual observer as a powerful and moving example of structural art.

In the mid-19th century New York City and Brooklyn were among America's largest commercial centers. Separated by the East River, the two cities required extensive use of ferries for transportation. As a bridge builder, John A. Roebling (1806–69) recognized the desirability of building a permanent crossing over the river, and in the late 1850s he developed plans for a 1,595-foot-span suspension bridge to connect the two cities. The scope of the project and coming of the Civil War, however, prevented any serious work from being done. After the war, Roebling's Cincinnati Suspension Bridge (1867, 1899) attracted considerable attention, and this success, in conjunction with the brutal winter of 1866–67 that severely disrupted ferry service in the New York Harbor, prompted formation of the New York Bridge Company. With sanction from the state legislature to build a bridge across the East River, the company appointed Roebling as its chief engineer in May 1867.

Pedestrian promenade on
top of the Brooklyn Bridge,
c. 1905.

During the next two years he developed final plans for the structure. Just as construction work began in late June

1869, however, a ferry boat pulling into a Brooklyn wharf crushed his left foot. Despite amputation of his foot, Roebling succumbed to tetanus on July 22. Rather than abandon the project, the bridge company selected his 32-year-old son, Washington Roebling (1837–1926), as the new chief engineer. He had already assisted his father in developing the design and had taken a trip to Europe to learn more about the pneumatic caisson technology recently developed there. During the next 14 years Washington Roebling, with the crucial assistance of his wife, Emily, oversaw construction of the masonry towers, the steel suspension cables, the two-tier traffic deck and all other ancillary features of the design.

As civil engineering historian Robert Vogel has noted, "The Brooklyn Bridge is a perfect example of a structure whose design is the result of logical evolution and proven construction methods, yet also embracing innovative technologies and materials. But even in the case of the new departure, nothing was adopted by either John or Washington Roebling that had not been shown to be effective and safe in prior undertakings." In particular, Roebling's Cincinnati Suspension Bridge and Niagara Railroad Suspension Bridge (1855) in Niagara Falls served as the most important precedents for the East River span.

Construction of the Brooklyn Bridge began with excavation for the tower foundations using pneumatic caisson techniques, and in 1870–71 workers started laying stonework on top of both wooden caissons. After more than five years both Gothic-style masonry towers were completed, and in the summer of 1876 the first wire rope physically connecting New York and Brooklyn was raised. Cable making and wrapping continued from 1877 through 1879. Not long afterward, erection of the sus-

Aerial view looking down on the Brooklyn Bridge's tower and cables.

Right: Tower, cable and deck of Roebling's Brooklyn Bridge, with the Manhattan Bridge in the background. Below: Under the main traffic deck.

pended deck began, followed by installation of the diagonal sway bracing. On May 23, 1883, official ceremonies to open the bridge were held, and later that summer cable railway service over the span started.

The bridge has a clear span of 1,595 feet between the towers and an overall length of more than 3,400 feet (excluding approaches). Each of the four main cables has a diameter of almost 16 inches and a total length of 3,578 feet from anchorage to anchorage. The masonry towers both rise 276 feet above high water, while the arches in the towers are 117 feet above the roadway. In addition, the navigational clearance at midspan is 135 feet above high water, a height sufficient to allow passage of both sailing ships and modern vessels.

Since its completion the Brooklyn Bridge has elicited a strong response from artists, poets and the public. But beyond its stature in American popular culture, the bridge stands as a major accomplishment in structural engineering. For many years it was the longest span bridge in the world and was not significantly surpassed until completion of the George Washington Bridge across the Hudson River in 1931. In the early 1950s famed engineer David B. Steinman (1886–1960) supervised a major refurbishment of the bridge that substantially altered the deck structure to accommodate automobile traffic better. But great care was taken to ensure that the basic appearance and profile of the bridge remained unchanged. Of most importance to visitors wishing to appreciate the glories of Roebling's design firsthand, the elevated pedestrian promenade extending across the bridge has been preserved and is open to the public. For a bridge lover, a walk across the span is the experience of a lifetime. NR, ASCE.

After designing world-famous Central Park in Manhattan, Frederick Law Olmsted (1822–1903) and Calvert Vaux (1824–95) began work on a similar project for Brooklyn in 1865. Although perhaps not as well known as its counterpart across the East River, Prospect Park is considered a major achievement in urban park design. Olmsted and Vaux designed numerous paths and trails for both horses and pedestrians, requiring the construction of many small-scale bridges. Seeking to build a variety of designs, Vaux selected a new form of concrete technology for one of these structures. As documented by William P. Chamberlin, the Cleft-Ridge Span is the first concrete arch bridge in the United States. It contains no iron or steel reinforcing and is an example of precast "artificial stone" construction popularized by, among others, French engineer François Coignet and therefore often referred to as "béton Coignet," or Coignet's concrete. In the early 1870s Gen. Quincy A. Gillmore, an artillery and fortifications expert of the U.S. Army Corps of Engineers, publicized the technology in his book *A Practical Treatise on Coignet Béton and Other Artificial Stone.*

The Cleft-Ridge Span is only 20 feet in length but has a width of more than 60 feet; in some respects, it is more like a tunnel than a bridge. Although the structure has deteriorated somewhat during the last 100 years, the detailed Gothic ornamentation cast into the surface of the artificial stone is clearly visible. Like most early precast artificial stone, béton Coignet had primarily a regional popularity (i.e., East Coast), but it served as an important precedent for the later use of concrete as a vehicle for ornamentation.

■ Cleft-Ridge Span

In Prospect Park
Near the Ocean Avenue
entrance
Calvert Vaux
1872

Above: Portal facade of the Cleft-Ridge Span in Prospect Park. Below: Detail of the ornamentation made possible using "béton Coignet."

■ George Washington Bridge
Across the Hudson River
On Interstate 95
Othmar H. Ammann
1931

After completion of the 1,595-foot-long Brooklyn Bridge in 1883, no suspension bridge exceeded it in length by more than a few hundred feet for more than 40 years. But in the late 1920s the newly formed Port Authority of New York began planning a huge highway bridge across the Hudson River between northern Manhattan and New Jersey. Under the supervision of Othmar H. Ammann, the Port Authority started construction of the 3,500-foot-long George Washington Bridge in 1927. Designed for vehicular traffic, the bridge included a two-tier deck with the top tier for automobiles and the bottom for rapid-transit rail lines. Since the completion of the bridge, the rapid-transit lines have been removed and both tiers now carry automobile traffic.

As noted by David Billington, professor of civil engineering at Princeton University, Ammann consciously considered the aesthetic appearance of the bridge and derived its basic design from early 19th-century British suspension bridges. Originally, Ammann planned to encase the structural steel towers in masonry, but the economic effect of the Depression prompted the Port Authority to abandon the masonry facing as a cost-cutting measure. As a result, the steel frame towers, which were never intended to be visible, have now become one of the most visually distinctive parts of the structure. After more than 50 years of service, the bridge continues to carry traffic on Interstate 95. ASCE.

Above: Cables anchored under the George Washington Bridge. Right: Underside of the bridge's deck looking toward Manhattan.

Left: George Washington Bridge towers, originally intended to be covered by masonry. Below: Hell Gate Bridge, c. 1920. The masonry towers serve no structural purpose.

As part of the construction of New York City's Pennsylvania Station, a new connecting railroad line was built to join the Pennsylvania Railroad with the New Haven Railroad. The two giant companies formed the New York Connecting Railroad, which subsequently built a huge steel through arch bridge over the East River's Hell Gate to complete the link between the two transporation systems. With a clear span of 977 feet between hinges, the Hell Gate Bridge is designed for a combined dead and live load of more than 75,000 pounds per linear foot along the deck. Visually, the design by Gustav Lindenthal (1850–1935) is distinguished by enormous masonry towers at both abutments. These serve no real structural purpose but were intended to lend an aura of graceful dignity to the massive steel bridge. Some connoisseurs of bridge design consider the Hell Gate span a masterpiece of bridge architecture, while others call the masonry towers a debasement of the principles that exemplify the best in structural art. The bridge continues to carry heavy railroad traffic on a daily basis.

■ **Hell Gate Bridge**
Across the East River
Adjacent to the
Triborough Bridge
Gustav Lindenthal
1917

High Bridge before its alteration in the 1930s.

■ High Bridge
Across the Harlem River
Adjacent to the
Washington Bridge
John B. Jervis
1842

The construction of New York City's Croton Aqueduct in the 1830s and early 1840s entailed building several masonry bridges to carry the city's new water supply across ravines and other natural obstacles. By far the largest barrier that lies between the Croton River and consumers in Manhattan is the Harlem River, a natural waterway that forms the northern boundary of Manhattan Island. John B. Jervis, the project engineer, could have built a pressurized siphon to carry the aqueduct under the river; however, as documented by historian Larry Lankton, he recognized the symbolic importance that a large masonry span could hold for the people of New York, who were financing the expensive aqueduct project. Consequently, he opted for a bridge rather than a siphon. The High Bridge rises more than 100 feet above the river and is slightly less than 1,200 feet long. It originally consisted of 15 masonry arches, seven of which had clear spans of 50 feet and eight of which had clear spans of 80 feet, and carried three wrought-iron pipelines built in the mid-19th century. In 1937 five arches over the river proper and the adjacent railroad tracks were replaced by a single steel-plate girder arch, which provided greater clearance for ships and barges. Although Jervis's original Croton Aqueduct is no longer used, the High Bridge still plays a role in the region's water supply system.

■ Manhattan Bridge
Across the East River
At Pike Street
O. F. Nichols
1909

About one-half mile upstream from the Brooklyn Bridge, the Manhattan Bridge provides another highway connection between Brooklyn and Manhattan. The third major suspension bridge across the East River, it has a main span between towers of 1,480 feet. The Manhattan Bridge differs from John A. Roebling's design in its use of steel rather than masonry towers and from the Williamsburg Bridge (1903) in its employment of a thin Warren truss for the deck instead of a bulky lattice truss. The cables and deck lend the Manhattan Bridge a decidedly modern appearance, but the unorthodox classicism of the steel towers belies the sophisticated engineering techniques involved in its construction. After almost 80 years, the bridge still remains in service. NR.

Queensboro Bridge shortly after completion, c. 1910.

Just as Brooklyn and Manhattan required permanent bridges to connect them across the East River, it also became important to join Manhattan with the borough of Queens. Located just to the north of Brooklyn, Queens did not develop as rapidly as its southern neighbor, but by the early 20th century it was large enough to warrant construction of its own link to Manhattan. Because the planned Queensboro Bridge extended across Blackwell's Island, now called Roosevelt Island, it was not necessary to construct a long-span suspension bridge. Instead, Gustav Lindenthal chose to build a massive cantilever bridge that used piers built on the island. The bridge has a total length of 1,182 feet, not counting approach spans, and is visually distinguished from simple truss bridges by having the steel truss increase in depth where it passes over the four interior piers. The Queensboro Bridge is the only major cantilever bridge in New York City, where it is known locally as the 59th Street Bridge, a name made famous in a popular song by Simon and Garfunkel. NR.

■ **Queensboro (Blackwell's Island) Bridge**
Across the East River
At 59th Street (State Route 25)
Gustav Lindenthal
1909

The first important project undertaken by Robert Moses's Triborough Bridge and Tunnel Authority (and the source of its name) was a three-armed complex of bridges and roads connecting the boroughs of Manhattan, Queens and the Bronx. The largest component of this system is the 1,380-foot-long, clear-span suspension bridge across the Hell Gate between Wards Island and

■ **Triborough Bridge**
Over the East River
On Interstate 278, at the Hell Gate
Othmar H. Ammann and Allston Dana
1936

Triborough Bridge, the first major civic monument of Robert Moses.

Queens. This structure has impressive steel towers that rise more than 300 feet above high water: the traffic deck is a riveted Warren truss that is well adapted to resisting aerodynamic (wind) forces. In contrast to the Brooklyn Bridge, however, the Triborough Bridge does not have diagonal stay cables connecting the deck and towers, which would also help counter wind stress. Although much smaller than Othmar Ammann's previously completed George Washington Bridge (1931), the Triborough Bridge represented an important step in the modern development of New York City and opened the way for increased highway traffic in the region. ASCE.

■ **Verrazano Narrows Bridge**
Across the Verrazano
Narrows
On Interstate 278
Ammann and Whitney
1964

Named after Giovanni da Verrazano, the Italian explorer who discovered New York Harbor in April 1524, the Verrazano Narrows Bridge is the second longest suspension bridge in the world after the Humber River Bridge in Great Britain. It stretches 4,260 feet from tower to tower and has a total length of 7,200 feet. To avoid impeding navigation in and out of the harbor, it also provides a clearance of 216 feet between the water level and the bottom of the bridge deck. Erected under the direction of Othmar H. Ammann, the bridge's design used computers to help ensure a safe, yet economical, structure. Highway engineers had long dreamed of connecting Staten Island with Brooklyn, and in the 1920s work began on a tunnel between the two boroughs. However, this project soon aborted. It was not until the late 1950s that the Triborough Bridge and Tunnel Authority began planning in earnest for a bridge. The proposal prompted considerable controversy, including the opposition of historian and design critic Lewis Mumford, largely because its approach lanes required the removal of 8,000 Brooklyn residents from their homes. Despite this, construction of the huge structure proceeded relatively rapidly, and it was opened to traffic in November 1964.

Verrazano Narrows Bridge, which looks like Othmar H. Ammann's Delaware Memorial Bridge, completed a few years earlier.

Washington Bridge, a multi-rib, plate-girder arch structure, c. 1905.

The northern end of Manhattan Island is separated from the Bronx by the Harlem River, and in the 1880s planning began for a high-level highway bridge that would carry traffic approximately 110 feet above the river. After the city considered a wide variety of designs, it selected a two-span, two-hinged steel deck arch supplemented by several masonry arch approach spans. The two main arches are 508 feet long and consist of riveted plate girders with a maximum depth of nine feet. Each of these main spans is actually made up of six parallel arches that carry the traffic loads to the structure's foundations. Because of the relatively deep and narrow valley that surrounds the Harlem River, the large deck arch design was particularly well suited to the site. In fact, the Washington Bridge is the only major steel deck arch bridge in the New York City area. NR.

■ **Washington Bridge**
Across the Harlem River
At 181st Street
William R. Hutton and
Edward H. Kendall
1889

Following completion of the Brooklyn Bridge in 1883, commerce between Manhattan and Brooklyn continued to accelerate. By the beginning of the 20th century, economic pressure prompted the construction of two more large bridges between the boroughs to meet increased traffic demands. The first of these was the Williamsburg Bridge, built slightly more than two miles upstream from John A. Roebling's landmark. Like the Brooklyn Bridge, it is a suspension span but has steel rather than masonry towers. It was also less expensive, built in less time and longer than its rival, becoming the world's longest suspension bridge when it opened. The deck truss of the Williamsburg Bridge is a bulky lattice structure with a depth of 40 feet. The massive dimensions of this truss give the 1,600-foot-long structure an awkward appearance that suffers in comparison with other, more graceful, 20th-century suspension bridges.

In the mid-1980s structural studies concluded that a continuing problem with the bridge's cables, woven of 16,325 miles of ungalvanized wire, required resolution by 1995. Deterioration and breaks in the cables, fabricated differently from the Brooklyn Bridge's system, have become so pervasive that the city is wrestling with what to do with the bridge. Two basic options include spending

■ **Williamsburg Bridge**
Across the East River
At Delancey Street
Leffert Lefferts Buck
1903

Left: Williamsburg Bridge, c. 1905. Right: Aerial view with north Brooklyn in the background.

$250 million to replace the span's cables, which has never been done while a bridge is in use, or building an entirely new structure at perhaps twice the cost. In either case, bridge-building or preservation history may be in the making.

NORTH BLENHEIM

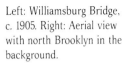

■ **Old Blenheim Bridge**
Across Schoharie Creek
Adjacent to State
Route 30
Nicholas Powers
1855

After the Bridgeport Covered Bridge (1862) in Grass Valley, Calif., the Old Blenheim Bridge is recognized as the longest single-span covered bridge in the world and a rare surviving example of a double-barreled wooden truss bridge. The latter distinction refers to the structure's use of three sets of trusses that provide for two separate traffic lanes. The bridge is a Long truss, patterned after designs developed by Stephen H. Long in the 1830s, with an auxiliary wooden arch attached to the center truss. It has a clear span between abutments of 210 feet. Local businessmen formed the Blenheim Bridge Company in the early 1850s to finance construction of the span. Responsibility for its erection fell to Nicholas Powers, a prominent covered-bridge builder from Vermont who was brought to North Blenheim specifically to design and supervise construction of the wooden superstructure. Powers's bridge remained in use until 1932, when a new steel truss was built nearby to carry highway traffic across Schoharie Creek. For more than 50 years local authorities have preserved the Old Blenheim Bridge as a historic site. NR, ASCE.

Old Blenheim Bridge when it still carried highway traffic, c. 1925

Downstream face of the Olive Bridge Dam under construction, c. 1915.

STONY HOLLOW

By the beginning of the 20th century, the city of New York was concerned that its water supply system would be insufficient to provide for the municipality's future growth. Even with completion of the long-planned New Croton Dam (1907), it appeared that the expansion of the city's Croton Aqueduct system would soon prove inadequate for projected needs. In 1905 the city formed a water supply board that began planning for a new dam and aqueduct system to store, deliver and distribute water for municipal use. After intensive study, this board selected the Catskill Mountains, 100 miles north of New York City, as the new source for the city's water supply system.

The primary storage reservoir for the Catskill Aqueduct is formed by the Olive Bridge Dam and two smaller earthen dikes. The main part of the dam consists of a 1,000-foot-long, straight-crested concrete gravity structure with a maximum height of more than 250 feet. It impounds Ashokan Reservoir, an artificial lake with a capacity of more than 132 billion gallons, enough water to cover Manhattan Island to a depth of 30 feet. Water from the reservoir flows by gravity (no pumping is necessary) to New York City via a 120-mile-long aqueduct. After entering the Catskill Aqueduct at the Olive Bridge Dam, the water takes approximately three days to reach Staten Island.

■ Olive Bridge Dam
Across Esopus Creek
Adjacent to State
Route 28A
J. Waldo Smith
1916

VALHALLA

With the expansion of New York City's water supply system into the Catskill Mountains in the early 20th century, the water supply board recognized the need to build a large storage and distribution reservoir near the city. Built on the site of a small 19th-century reservoir, the Kensico Dam is a straight-crested concrete gravity structure with a maximum height of more than 300 feet and a length of 1,025 feet (not counting the shallow earthen dikes). Although the dam consists primarily of concrete, its downstream side is faced with masonry quarried near the dam site. The masonry facing is structurally unnecessary, but the city believed that this

■ Kensico Dam
Across the Bronx River
Adjacent to State
Route 144
J. Waldo Smith
1916

Ornamentation on the downstream face of the Kensico Dam.

architectural treatment lent the dam a monumental dignity befitting its municipal importance. The 22 masonry panels visible on the downstream face are separated by expansion joints built into the concrete to prevent shrinkage cracks. A 75-mile-long aqueduct connected to the Ashokan Reservoir in the Catskills delivers water to the Kensico Reservoir, which then supplies it to New York City.

■ ■ ■ ■ ■ ■ PENNSYLVANIA ■ ■ ■ ■ ■ ■

BROWNSVILLE

■ **Dunlap's Creek Bridge**
Across Dunlap's Creek
On old U.S. Route 40
Richard Delafield
1839

Left: Detail of the original arch used in the Dunlap's Creek Bridge, with some recent additions. Right: Postcard depiction of the bridge, c. 1909.

In the early 19th century the National Road leading west out of Maryland incorporated numerous stone arch bridges into its right-of-way. In addition, the road prompted construction of the first iron arch bridge in the United States. Built by the U.S. Army Corps of Engineers, the Dunlap's Creek Bridge remained a structural anomaly for many years, for it was not until the late 19th century that metal arch highway bridges received even a modest amount of acceptance among American engineers. The structure spans 80 feet between abutments and consists of five parallel tubular ribs. Each rib is made of nine cast-iron elliptical segments bolted together to form an 80-foot-long arch. Today, the bridge is still in use, but 20th-century development in the area has largely obscured the historic structure from view. NR, ASCE.

COLUMBIA

During the early years of westward migration, the Susquehanna River formed a major impediment to transportation. For many years a ferry operated at the site of the future Old Columbia–Wrightsville Bridge, and in 1812 the first of four bridges that preceded the present structure came into service. The first three of these were wooden covered bridges, while the fourth was a steel Pratt through truss. With the growth of automobile traffic in the 1920s, state highway authorities concluded that it would be desirable to replace the steel truss span between Columbia and Wrightsville with a newer, wider concrete structure. Interest in the project was heightened because the bridge was part of the Lincoln Highway (U.S. Route 30) connecting Philadelphia and Pittsburgh. The new reinforced-concrete bridge incorporated 28 three-ribbed, open-spandrel deck arches into a design with a total length of more than 6,600 feet. Each of the 28 arches has a span of 185 feet. The bridge no longer carries modern U.S. Route 30, but it is still an important part of the state's transportation system. ASCE.

■ **Old Columbia–Wrightsville Bridge**
Across the Susquehanna River
On State Route 462, between Columbia and Wrightsville
James B. Long
1930

HARRISBURG

In the late 19th century the Pennsylvania Railroad chose to invest in building several large masonry arch bridges that would not require extensive maintenance expenditures or replacement anytime in the foreseeable future. Such bridges were expensive but the railroad could easily afford to finance them. The largest and most impressive of these structures is the Rockville Bridge north of Harrisburg. It is 3,820 feet long and consists of 48 arches, each spanning 70 feet. The massive structure is still in active use. NR, ASCE.

■ **Rockville Arch Bridge**
Across the Susquehanna River
North of Interstate 81
William Hood
1902

Two views of the Rockville Arch Bridge crossing the Susquehanna River north of Harrisburg.

■ Walnut Street Bridge

Across the Susquehanna
River at Walnut Street
Phoenix Bridge Company
1890

Left: Walnut Street Bridge,
Harrisburg. Right:
Wallenpaupack Dam, Hawley.

Built for vehicular traffic across the Susquehanna River,
the Walnut Street Bridge is a superb example of a pin-
connected Baltimore through truss. It is a 14-span
structure with an overall length of more than 2,800 feet.
Eleven of the spans are 175 feet long, and three are 240
feet long. In addition to its subdivided panels, the bridge
is distinguished by its use of cylindrical Phoenix columns
for all the main compression members. Although no
longer called on to carry highway traffic, the bridge is still
maintained for bicycle and pedestrian use and provides
ready access to City Island in the middle of the river. NR.

HAWLEY

■ Wallenpaupack Dam

Across Wallenpaupack Creek
Adjacent to U.S. Route 6
Electric Bond and Share
Company
1926

In the 1920s the Pennsylvania Power and Light Company
expanded its service area through the consolidation of
many smaller companies in the region. The Wallenpaupack Hydroelectric Plant, located near the Pocono
Mountains, was designed to supplement the generating
capacity of the firm's numerous steam plants by providing power during times of peak load. By allowing the
entire system to function at a higher load factor, the new
plant greatly enhanced the company's economic stability.
The Wallenpaupack Dam is a 64-foot-high, 1,280-foot-
long concrete and earthen gravity dam that forms a
storage reservoir more than 13 miles long with an average
width of one mile. Many visitors to the resort area assume
that Lake Wallenpaupack is primarily a recreation
facility. Its technological purpose is, however, much more
important in that it provides electricity to hundreds of
thousands of residents in the Philadelphia area.

HOLTWOOD

■ Holtwood (McCall Ferry) Dam

Across the Susquehanna
River
1 mile west of State
Route 72
Hugh L. Cooper
1910

Located on the lower Susquehanna River only a few miles
from the Pennsylvania-Maryland border, the Holtwood
Dam is a 59-foot-high, 2,392-foot-long concrete overflow
gravity dam. In 1906 the McCall Ferry Power Company
began construction of the huge dam as part of a
hydroelectric power project. Work was suspended a year
later during the financial panic of 1907 when financing
for the structure collapsed. The property eventually came
under the control of the Pennsylvania Power and Light
Company, which completed work on the dam and
brought the power plant on line in 1910. At that time it was

one of the largest hydroelectric generating facilities in the United States, and it helped secure for Hugh L. Cooper an international reputation as an expert in hydroelectric design. The dam still functions as a major component in the company's electric power system.

KUSHEQUA

In the early 1880s the Erie Railroad built a new line to serve the mining region of northwestern Pennsylvania. This line required a high-level crossing of Kinzua Creek. Built by the Phoenix Bridge Company, the 2,053-foot-long, 301-foot-high, wrought-iron Kinzua Viaduct started carrying traffic in 1882. Less than 20 years later the Erie decided to replace the spindly original structure with a stronger steel design capable of supporting heavier train loads. This new viaduct remained in operation for much of the 20th century before the Erie abandoned the entire Bradford Division line. Today, the structure stands unused in the Kinzua Bridge State Park. ASCE.

■ Kinzua Viaduct
Across Kinzua Creek Valley
On the former Erie
Railroad–Bradford Division
line, 5 miles south of State
Route 59
Octave Chanute and Mason
Strong
1900

Above: Kinzua Viaduct after its reconstruction, c. 1905. Left: Construction view of the Holtwood (McCall Ferry) Dam, 1907.

John Roebling's Delaware
Aqueduct from the New York
shore.

LACKAWAXEN

■ **Delaware Aqueduct**
Across the Delaware River
Near State Route 590,
between Lackawaxen, Pa.,
and Minisink Ford, N.Y.
John A. Roebling
1849

Most famous for his work in designing the Brooklyn Bridge (1883), John A. Roebling built many suspension bridges before receiving the commission for a span across the East River in New York City. The oldest surviving Roebling suspension bridge is the Delaware Aqueduct across the Delaware River on the New York –Pennsylvania border. It was built by the Delaware and Hudson Canal Company, a shipper of anthracite coal and a forerunner of the Delaware and Hudson Railroad, as one of four aqueduct bridges on a canal route connecting Honesdale, Pa., and Kingston, N.Y. The iron cables of the suspension span supported a wooden trough filled with water that allowed passage of canal boats over the river. In the early 20th century the canal ceased operation, and the bridge was adapted for vehicular traffic. Now owned by the National Park Service, the bridge has been restored recently and is readily accessible to the public. NR, ASCE.

LANESBORO

■ **Starrucca Viaduct**
Across Starrucca Creek
On the Erie Railroad right-
of-way, near State Route 71
John P. Kirkwood
1848

With the expansion of America's railroad system in the 1840s, lines began spreading westward toward the newly settled frontier. By this time New York City was one of the country's most important commercial centers, and competing railroads vied to serve the burgeoning metropolitan region. Among the largest and most ambitious of these was the New York and Erie Railroad, a British-backed company with access to considerable capital reserves. Designed to connect New York with the Great Lakes region (hence the name Erie), the railroad's main line extended through northern Pennsylvania and southern New York, crossing several valleys.

The most imposing of these was the valley drained by Starrucca Creek, a tributary of the Susquehanna River. Choosing to emulate the permanent and expensive character of British railroads, the company decided to build a massive stone viaduct. Constructed over a two-year period, the viaduct is 1,040 feet long and incorporates 17 arches into its design, each with a span of 50 feet. The tracks are a maximum of 100 feet above the founda-

Starrucca Viaduct, Lanesboro. Left: Detail of the masonry piers. Below: Aerial view.

tions, and the piers are 12 feet thick at the bottom. The viaduct cost more than $300,000 and almost forced the company into bankruptcy. The quality of the design and the integrity of construction proved to be remarkably good, however, and the structure can still carry heavy railroad traffic. ASCE.

NICHOLSON

This massive reinforced-concrete arch bridge is usually considered to be volumetrically the largest structure of its type in the world. It is 2,230 feet long and consists primarily of 10 semicircular, two-ribbed, open-spandrel arches with spans of 180 feet. The bridge has a maximum height of 240 feet above the stream bed. As engineering historian Carl Condit has pointed out, however, the pier foundations extend another 100 feet below the surface to reach bedrock. The viaduct is still used to carry railroad traffic. ASCE.

■ **Tunkhannock Creek Viaduct**
Across Tunkhannock Creek
On the Delaware, Lackawanna and Western Railroad right-of-way, near State Route 92
George J. Ray
1915

Massive concrete arches of the Tunkhannock Creek Viaduct.

NORTH VERSAILLES

■ **George Westinghouse Memorial Bridge**
Across Turtle Creek
On U.S. Route 30
Vernon R. Covell
1931

Named in honor of the electrical engineer who founded the world-renowned Westinghouse Company, this structure is among America's largest long-span, reinforced-concrete highway bridges. Built to carry the historic Lincoln Highway across Turtle Creek Valley, the Westinghouse Bridge consists of five two-ribbed, open-spandrel arches, the largest of which is in the center of the bridge and spans 460 feet. This parabolic arch has a particularly striking appearance because it rises more than 150 feet from the foundation to the crown. The bridge's name is particularly appropriate because of the structure's close proximity to one of the Westinghouse Company's oldest factory complexes. NR.

PHILADELPHIA

■ **Ben Franklin Bridge**
Across the Delaware River
On Interstate 76
Ralph Modjeski and
Leon Moisseiff
1926

For a short time the Ben Franklin Bridge was the longest suspension bridge in the world. Spanning 1,750 feet between towers, the bridge provided access between Philadelphia and Camden, N.J., without impeding river traffic that serves the region's international port facilities. Used today to carry both vehicular traffic and a rapid-transit rail line, the Ben Franklin Bridge recently underwent major refurbishment to ensure its serviceability well into the 21st century. It stands as a major visual landmark in a city brimming with historic sites.

■ **Fairmount Dam**
Across the Schuylkill River
On East Drive at Fairmount
Waterworks
Schuylkill Navigation
Company and Philadelphia
Water Company
1821

The city of Philadelphia is located at the confluence of the Delaware and Schuylkill rivers. Draining a substantial portion of southeastern Pennsylvania, the Schuylkill flows from the Pottsville-Reading region before passing through Philadelphia on its way to the Atlantic Ocean. In the early 19th century the Schuylkill Navigation Company constructed a canal to transport coal and other goods along the river. As part of its transportation system the company built several timber crib dams to raise the level of the river and facilitate slackwater navigation. At the same time, the city of Philadelphia was developing the Schuylkill as a source of domestic water supply. In 1801 Benjamin H. Latrobe (1764–1820) completed a steam-powered pumping plant — the original Fairmount Waterworks — to serve the city, and in 1819 the city, in concert with the Schuylkill Navigation Company, began work on a timber crib dam across the river both to aid navigation and to allow for water-powered pumping from the new Fairmount Waterworks. As documented by Philadelphia historian Jane Mork Gibson, the 30-foot-high dam stretched obliquely across the river with a total length of slightly more than 1,200 feet. It was erected under the supervision of Capt. Ariel Cooley and first carried water over its crest in July 1821. During the 19th century the original structure was repaired and strengthened on numerous occasions, but it continued to function as part of the city's water supply system. In 1911 the city decommissioned the Fairmount Waterworks but the dam was left in place. Today, parts of the waterworks are being

Above: Westinghouse Bridge, with its prominent reinforced-concrete arches, c. 1940. Left: Philadelphia anchorage of the Ben Franklin Bridge, c. 1920. Below: Fairmount Dam as it appears today.

studied for reuse as a hydroelectric power plant, and the Fairmount Dam may once again provide a public service to the residents of Philadelphia. NR.

■ Frankford Avenue Bridge
Across Pennypack Creek
On Frankford Avenue
1697, 1893

Following the establishment of William Penn's colony in 1681, the region surrounding Philadelphia began to grow rapidly with the influx of new settlers. In the late 1690s work on the King's Road between Philadelphia and New York City began and included the construction of several bridges, one of which was a three-span, 75-foot-long, stone arch structure across Pennypack Creek. Although strengthened and widened for trolley service in the 1890s, the bridge still retains much of its original design. It is considered the oldest bridge in the United States that continues to serve as part of a modern highway system. ASCE.

Frankford Avenue Bridge after strengthening with auxiliary reinforced-concrete arches.

■ Walnut Lane Bridge
Across Lincoln Drive
On Walnut Lane
Philadelphia Department of
Public Works
1950

In reinforced-concrete girder bridges, the steel reinforcing bars are placed in the lower part of the girders in order to withstand tensile stresses. Usually the reinforcing bars are simply imbedded in the structure during construction and placed in tension because of the force of gravity acting upon the girder and any live loads placed upon it. However, it is possible to reduce the amount of material in such a girder by "prestressing" (stretching) the steel reinforcing members so that they are in tension before the placement of any heavy loads on the girder. The Walnut Lane Bridge is the first prestressed reinforced-concrete bridge built in the United States. It is a girder bridge with a main span of 160 feet and two approach spans of approximately 50 feet. The steel reinforcement for each girder in the main span consists of four steel cables that are prestressed in order to carry traffic loadings more efficiently. The main girders in the Walnut Lane Bridge require a depth of only six feet, seven inches; without prestressing, the girders would require much greater depth and hence much more material. In the wake of the technology's success at Walnut Lane, prestressed concrete has been used in numerous highway bridges and buildings throughout the United States.

Panther Hollow Bridge in its park setting.

PITTSBURGH

Because of its location, straddling the Allegheny and Monongahela rivers, Pittsburgh came to be known as the City of Bridges and it can still lay claim to this title. Because of extensive replacement efforts in recent years, however, the vast majority of these structures postdate World War II. Among the most picturesque historic spans remaining in the City of Bridges is the Panther Hollow Bridge. Built in the 1890s, the single-span, three-hinged steel deck arch bridge is 360 feet long and rises approximately 120 feet above the stream bed. The floor system is supported on four ribs spaced 12 feet apart. The graceful curve of the arches was considered particularly appropriate for the structure's park setting.

The lenticular truss is an unusual type of bridge that is distinguished by the polygonal configuration of its top and bottom chords. The name of the truss is derived from the lenslike shape formed by the top and bottom chords. Although most lenticular trusses in the United States were built by the Berlin Iron Bridge Company, the Smithfield Street Bridge is a superb example of the technology designed by the prominent engineer Gustav Lindenthal. In fact, it represents his first major design commission.

length of 364 feet. When originally built, the structure incorporated only two parallel sets of trusses into its design. In 1891 a third set of trusses was added to allow the complete separation of vehicular traffic from a trolley line. Since that time the bridge has been continually maintained and refurbished, but its appearance has generally remained unchanged. The one exception to this came in 1915 when the original masonry, mansard-roof portal structures were replaced with the existing cast-

■ **Panther Hollow Bridge**
Across Panther Hollow
In Schenley Park, on Panther Hollow Road
H. B. Rust
1896

■ **Smithfield Street Bridge**
Across the Monongahela River
On Smithfield Street
Gustav Lindenthal
1883, 1891

Smithfield Street Bridge, Pittsburgh's famous lenticular truss span.

steel design. The Smithfield Street Bridge stands as the city's last great river span surviving from the 19th century and holds a special place in Pittsburgh's cultural heritage. NR, ASCE.

SUMMIT

■ **Wells Creek Bridge**
Across the Baltimore and Ohio Railroad right-of-way
On State Route 219
W. Bollman and Company
1871

Wendel Bollman is usually associated with the Bollman truss, but his company also manufactured other types of metal truss bridges when he perceived a market for them. The Wells Creek Bridge provides good evidence of his entrepreneurial skills in the bridge-building business. It is a single-span, pin-connected Warren through truss with an overall length of about 80 feet. The diagonal members of the truss are both cast iron (for the compression members) and wrought iron (for the tension members). Still owned by the Baltimore and Ohio Railroad, it is a rare surviving example of a pin-connected Warren truss.

WEST VINCENT

■ **Hall's Sheeder Bridge**
Across French Creek
On Hollow Road
Robert Russell and Jacob Fox, builders
1850

Although born in Connecticut, Theodore Burr (1771–1822) is closely associated with the state of Pennsylvania because of his career as a covered-bridge builder. In the early 19th century he constructed some of the world's largest covered bridges across the Susquehanna River at Harrisburg, Northumberland and

Hall's Sheeder Bridge, strengthened with a concrete pier at midspan.

McCall Ferry. But aside from major individual bridges, he is perhaps best remembered for his development of the Burr arch truss. This structural form used a multiple king-post truss, in which an arch is superimposed on the truss proper. Patented by Burr in 1817, the truss enjoyed wide popularity in the 19th century because it helped reduce the deflection of covered bridges at their midspan. The 100-foot-long Hall's Sheeder Bridge is a good example of a Burr arch truss still surviving in Burr's adopted state. The oldest wooden span in Chester County, it is still used to carry local highway traffic. NR.

■ ■ ■ ■ ■ ■ WEST VIRGINIA ■ ■ ■ ■ ■ ■

BARRACKVILLE

Characterized by extensive forests and mountainous terrain, West Virginia supported the construction of numerous 19th-century wooden covered bridges. Unfortunately, fires and floods have taken their toll on these venerable structures and only a handful survive. Completed in 1852 by the famous bridge builder Lemuel Chenoweth with the assistance of his brother Eli, the Barrackville Bridge is a 145-foot-long, single-span Burr arch truss. Built across Buffalo Creek, a small tributary of the Monongahela River located near the city of Fairmont in the north-central part of the state, it has been characterized by Richard S. Allen as an "outstanding structure" representing the best of West Virginia's long tradition of covered-bridge building. NR.

■ Barrackville Covered Bridge
Across Buffalo Creek
Lemuel Chenoweth and Eli Chenoweth, builders
1852

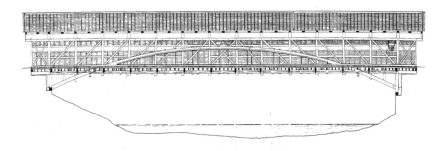

Above: Cross-sectional drawing showing the Burr arch truss design of the Barrackville Covered Bridge. Left: Contemporary view of the 145-foot-long span.

Above: Downstream side of the Lake Lynn Dam showing the overflow spillway. Right: Turbine-generator units in the Lake Lynn powerhouse.

CHEAT NECK

Constructed by the West Virginia Power and Transmission Company, the Lake Lynn Dam is used to provide hydroelectric power for northern West Virginia and southwestern Pennsylvania. A 135-high, 1,000-foot-long, concrete gravity overflow structure, it is equipped with movable tainter gates to provide for the release of heavy springtime floods. Located on top of a spillway, tainter gates are hinged steel structures that can be lifted or dropped to control the water level in a reservoir. The Lake Lynn Dam is now owned by the West Penn Power Company and remains in active use. Because of its close proximity to the West Virginia–Pennsylvania border, it was originally called the State Line Dam.

■ **Lake Lynn Dam**
Across the Cheat River
2 miles east of U.S. Route 119, at the West Virginia–Pennsylvania border
Sanderson and Porter
1926

MARTINSBURG

In 1832 the Berkeley County Court authorized funding for a bridge across Opequon Creek near the site of an old ford. The three-span, 165-foot-long masonry arch bridge was built to serve highway traffic along the major road between Martinsburg and Warm Springs, now known as Berkeley Springs. The uncoursed, ashlar limestone bridge has weathered the past 150 years in remarkably good condition and, aside from paving of the road surface, is essentially unchanged. It continues to handle a small amount of local traffic. NR.

■ **Van Metre Ford Stone Bridge**
Across Opequon Creek
On State Route 36 (Old Warm Springs Road)
Silas Harry, builder
1832

Van Metre Ford Stone Bridge, Martinsburg.

MILTON

In the early 1870s a line of the Baltimore and Ohio Railroad passed through Cabell County, in the southeastern part of West Virginia. As a result, the community of Milton achieved a degree of commercial prominence in the area and soon sought a permanent highway crossing over the Mud River. In late 1874 the county approved a covered wooden bridge in Milton, and by late 1875 the structure was in use. The Mud River Bridge is a single-span, 112-foot-long Howe truss with auxiliary arches. These arches were part of the original structure and are not a later addition. In the early 1970s the structure underwent extensive renovation, and some steel members were added to ensure the safety of the floor system. The bridge retains much of its structural integrity, however, and appears much as it did in the 1870s. It is still used to carry light vehicular traffic. NR.

■ **Mud River Covered Bridge**
Across the Mud River
On County Road 25, near U.S. Route 60
1875

PHILIPPI

■ **Philippi Covered Bridge**
Across the Tygart's Valley River
At U.S. Route 250
Lemuel Chenoweth, builder
1852

Before 1863 the present state of West Virginia was a mountainous and sparsely populated region of the Commonwealth of Virginia. With the founding of Virginia's Board of Public Works in 1816, several turnpike companies received state support to build and maintain highways through the region to the Ohio River watershed. These turnpikes required numerous bridges, and wooden structures were often built, largely because of the ready availability of timber. Lemuel Chenoweth, recognized as the most prominent and prolific builder of covered bridges in West Virginia, erected structures on several major roads, including the Staunton and Parkersburg Turnpike. In 1852, as part of the Beverly and Fairmont Road, he constructed the Philippi Covered Bridge, a two-span, 276-foot-long, double-barrel Burr arch truss. In recent years two additional concrete piers have been placed under each span to strengthen the structure. As a result, it is still capable of carrying heavy highway traffic on U.S. Route 250.

Merely because of its size and age, the Philippi Bridge warrants historical interest. But it is also noteworthy as the site of one of the first battles of the Civil War, a skirmish between Union and Confederate forces on the night of June 2, 1861. Given how easy it was to burn covered bridges and the strategic location of the Philippi Bridge, it is remarkable that the structure survived the war. NR.

RIPLEY

■ **Staats Mill Covered Bridge**
In Cedar Lakes Park
Off Jackson County
Route 25
H. T. Hartley, builder
1887; reerected 1986

The Staats Mill Covered Bridge is a late example of the Long truss patented by Stephen Long in 1830. Completed many years after the wooden Long truss achieved a modicum of success in the northeastern United States, this bridge is something of a historical anomaly. In 1971 the 97-foot-long bridge underwent a major overhaul, and its structural system was supplemented by large steel

Opposite: Philippi Covered Bridge, a survivor of the Civil War despite the flammability of its wooden design. Above: Interior of the Philippi Bridge. Left: Staats Mill Covered Bridge before being moved to escape flooding from a new reservoir.

girders under the traffic deck. The structural significance of the wooden truss then became relatively minor, but the bridge still retained much of its visual integrity. Further changes occurred in 1986, when it was moved from a reservoir impoundment area slated to be flooded. Now spanning a finger lake, the bridge's new home is in Cedar Lakes Park, a complex maintained as a camp by the state agricultural agency. NR.

SHEPHERDSTOWN

■ **Dam No. 4**
Hydroelectric Plant
Across the Potomac River
2 miles downstream from
State Route 480
Chesapeake and Ohio Canal
Company
1860

Downstream side of Dam
No. 4, extending straight
across the upper Potomac
River.

In the 1830s the Chesapeake and Ohio Canal Company built several wooden timber dams to provide water for the operation of its lock and canal system. In 1860 the company replaced the old dam near Shepherdstown, known as Dam No. 4, with a straight-crested, stone gravity overflow structure 22 feet high and 715 feet long. In subsequent years Dam No. 4 experienced some damage from floods but still remained operational. In 1906 the Martinsburg Power Company contracted with the canal company to build a hydroelectric power plant on the southern end of the dam. This plant came on line in 1909 and was purchased by the Potomac Edison Power Company in 1922. The plant and Dam No. 4 are still providing service to electric power consumers in the region. NR.

Original masonry tower supporting the Wheeling Suspension Bridge.

WHEELING

Completed in 1849 and rehabilitated several times in the 19th century, the Wheeling Suspension Bridge is a major monument in engineering history. Emory Kemp, America's leading scholar in the history of suspension bridges, has asserted that it is "the nation's most significant extant antebellum engineering structure" and its erection established "America's preeminent postion as a leader in the design and construction of long span suspension bridges." Designed and built by Charles Ellet (1810–62), who competed against John A. Roebling for the commission, it was an important link in the National Road, later U.S. Route 40. Because of navigation along the Ohio River, Ellet designed the structure with towers 1,010 feet apart, and for several years it boasted the longest clear span in the world.

In 1854 a windstorm caused considerable damage to the structure, and it was temporarily rebuilt as a 14-foot-wide bridge. William McComas, a former associate of Ellet, rebuilt the structure in 1859 with a slightly increased width. In 1872 Wilhelm Hildenbrand, an engineer with Roebling's company, reworked the deck structure and added diagonal stay wires between the towers and the deck to increase resistance to wind stresses. Finally, in the early 1980s, the West Virginia Department of Highways undertook a major project to rehabilitate the structure. Despite all this work, the bridge still bears a strong resemblance to Ellet's original design. NR, ASCE.

■ **Wheeling Suspension Bridge**
Across the Ohio River
On 10th Street, between Wheeling and Channel Island
Charles Ellet
1849, 1854, 1859, 1872

SOUTH

Norris Dam (1936) across the Clinch River in Tennessee. This concrete gravity structure, with its dramatic design by Roland Wank, was the first dam built by the TVA, one of the most famous programs of Franklin Roosevelt's New Deal.

CHILDERSBURG

■ **Kymulga Mill**
Covered Bridge
Across Talladega Creek
Adjacent to County
Road 46
c. 1870

Bridges were often built to provide access to regional commercial centers. Conversely, small communities, especially in rural areas, would at times develop in close proximity to an existing bridge. The Kymulga Mill Covered Bridge is located next to a small grist mill built in the 1860s to serve local farmers. Although bridge-mill complexes were a common phenomenon in 19th-century economic development, the Kymulga mill and covered bridge is the only site of its type to survive in Alabama with an intact covered bridge. The Town lattice truss is 105 feet long and is supported on masonry piers. It no longer carries highway traffic but is accessible to pedestrians visiting the nearby restored mill. NR.

ELGIN

■ **Wheeler Dam**
Across the Tennessee River
On State Route 101,
south of Elgin
U.S. Bureau of Reclamation
1936

Shortly after his inauguration in 1933, Franklin Roosevelt authorized the establishment of the Tennessee Valley Authority. The purpose of this new federal agency was to provide much-needed relief for the economically depressed region encompassed by the Tennessee River watershed. Dedicated to a diverse range of functions, including flood control, navigation, soil conservation, recreation and, perhaps most important, electric power generation, the TVA is best known for its large storage dams. The Wheeler Dam was the first dam built by the TVA across the Tennessee River proper. Located at the edge of the flood pool formed by the Wilson Dam (1925) near Florence, the Wheeler Dam is a 72-foot-high, 6,342-foot-long, straight-crested, concrete overflow gravity structure. At the southern edge of the dam is an outdoor hydroelectric generating plant with a capacity of more than 250,000 kilowatts. As with other TVA dams on the Tennessee River, navigation around the site is made possible by large canal locks.

MUSCLE SHOALS

■ **Wilson Dam**
Across the Tennessee River
On Colbert County Route 40
U.S. Army Corps of
Engineers
1925

In anticipation of America's entry into World War I, the 1916 National Defense Act authorized construction of a hydroelectric power dam on the Tennessee River at Muscle Shoals. The dam was intended to supply power for a nitrate plant that would boost America's capacity for munitions and explosives production. Construction of the Wilson Dam began in early 1918 but it was nowhere near completion at the time of Armistice Day in November 1918. In fact, it was not completed for another seven years.

The concrete dam is more than 4,800 feet long and has a maximum height of 137 feet. It is a straight-crested, overflow gravity structure with a generating facility built directly adjacent to the downstream side of the dam. In the 1920s the dam became the object of a major political fracas when Henry Ford attempted to purchase it for only

Left: Kymulga Mill Covered Bridge, with a nearby mill in the foreground. Below: Wheeler Dam, with its outdoor hydroelectric generating plant. Bottom: Neoclassical facade of the Wilson Dam powerhouse.

Main section of the Wilson Dam across the Tennessee River in northern Alabama.

a small fraction of its construction cost. The structure remained in public hands. however. and in the 1930s it came under the control of the Tennessee Valley Authority. Today, the dam has a generating capacity of more than 400,000 kilowatts and creates a large flood pool that facilitates navigation on the Tennessee River. NR.

ONEONTA

■ **Horton Mill
Covered Bridge**
Across the Calvert Prong of the Warrior River
Adjacent to State Route 75
Zelma C. Tidwell
c. 1930

Blount County is a hilly region of north-central Alabama that fostered construction of large covered bridges long after most jurisdictions considered them passé. In the late 1920s and early 1930s county officials authorized Zelma C. Tidwell and his uncle, Forrest Tidwell, to build at least four Town lattice trusses as part of the local highway system. One of these, the Horton Mill Covered Bridge, is an impressive two-span structure with an overall length of 220 feet. The bridge, still in local use, is preserved as a historic site and is readily accessible to visitors. NR.

Horton Mill Covered Bridge, a 20th-century Town lattice truss.

■ ■ ■ ■ ■ ■ ■ ARKANSAS ■ ■ ■ ■ ■ ■ ■

BENTON

The Old River Bridge is a 260-foot-long, two-span, pin-connected Pratt through truss supported on concrete-filled, steel cylindrical columns. Fabricated by an unknown bridge company, the structure is representative of thousands of Pratt trusses built in the United States during the late 19th and early 20th centuries. It was built at the Military Road Crossing of the Saline River, a transportation nexus established in the 1830s. Among the oldest surviving spans in the state, the Old River Bridge no longer carries highway traffic but is accessible to pedestrians. NR.

■ Old River Bridge
Across the Saline River
At River Road
1889

Old River Bridge, an excellent example of a 19th-century pin-connected Pratt through truss.

COTTER

A unique six-span, concrete rainbow arch design, the Cotter Bridge has five 216-foot arch spans, each reaching a height of 70 feet above the springline, and a smaller 130-foot span that crosses the Missouri Pacific Railroad line. It is among the largest rainbow arch bridges ever designed by the Marsh Engineering Company. Constructed under the supervision of Hal W. Hunt, chief engineer for the Bateman Contracting Company, the bridge was built using an innovative cableway method of erection that permitted construction without the placement of falsework or equipment in the river. The Cotter Bridge replaced a ferry and helped open up an undeveloped region in the Ozarks for social and economic development. It remains in service on a major regional highway. NR, ASCE.

■ Cotter Bridge
Across the White River
On U.S. Route 2
Marsh Engineering Company
1930

Cotter Bridge in the Arkansas countryside.

Wire cable and wooden deck of the Winkley Bridge in north-central Arkansas.

HEBER SPRINGS

■ **Winkley Bridge**
Across the Little
Red River
Adjacent to State Route 110
Harry Churchill, builder
1912

The best-known 20th-century American suspension bridges are large structures built across major rivers. But suspension bridges were also constructed in many rural areas to provide economical long-span crossings. The 550-foot-long Winkley Bridge is an excellent example of the latter type. Built by Harry Churchill, a local builder, between May and November of 1912, the bridge served local highway needs for 70 years before being closed in 1982. Since that time, it has been confined to pedestrian traffic and has become a popular tourist attraction in the Greers Ferry Lake region. NR.

SPRINGFIELD

■ **Springfield Bridge**
Across Cadron Creek
On Springfield Road (County Road 222), 2½ miles east of Springfield
King Wrought Iron Bridge Company
1874

The 188-foot-long Springfield Bridge is the oldest highway bridge and the only documented wrought-iron bowstring arch bridge in Arkansas. Based on Zenas King's 1867 patent and manufactured by his short-lived subsidiary in Kansas, the bridge has a main span of 146 feet, making it among the longest surviving bowstring arch bridges in the nation. Its single-span arch reaches a height of more than 15 feet from the bottom chord and is supported on two masonry stone piers. It remains today a significant example of the varied bridge designs supplied by private bridge manufacturing companies of the late 19th century. NR.

WAR EAGLE

■ **War Eagle Bridge**
Across War Eagle Creek
On County Road 98
Illinois Steel Bridge Company
1907

The War Eagle Bridge is a 182-foot-long, pin-connected Parker through truss built at the site of a river ford used from the 1830s through the early 20th century. This was also the site of an important grist mill, and by 1907 local commercial pressure prompted Benton County to build a permanent steel bridge at War Eagle. It is among the longest and oldest pin-connected Parker trusses still in use in America's highway system. NR.

Left: Springfield's bowstring arch truss. Below: Pastoral setting of the Springfield span. Bottom: Single-span War Eagle Bridge, a Parker truss.

■ ■ ■ ■ ■ ■ ■   FLORIDA   ■ ■ ■ ■ ■ ■ ■

MARATHON

**■ Seven Mile
(Knight's Key) Bridge**
Across the Gulf of Mexico
Adjacent to U.S. Route 1,
between Knight's Key and
Little Duck Key
Joseph C. Merideth
1912, 1938

In 1904 the Florida East Coast Railroad reached the town of Homestead, about 25 miles south of Miami. Under the direction of its president, Henry Morrison Flagler, the railroad sought to develop a southern terminus convenient for shipping through the Panama Canal, then under construction. After rejecting the idea of a lengthy extension across the mangrove swamp of the Florida Everglades, Flagler chose to build a 128-mile-long railroad from Homestead to Key West. This line was planned to run along the length of the Florida Keys and provide a permanent connection between Key West and the mainland. Much of the Key West Extension was built directly on the numerous keys, or islands, but several bridges were required to extend over significant stretches of water.

The longest bridge in the system is the Seven Mile Bridge, located about 45 miles east of Key West. Originally known as Knight's Key Bridge, this structure

Right: Aerial view of Knight's Key Bridge shortly after conversion to highway use, c. 1940. Below: Three-part postcard documenting the bridge's original construction, c. 1912.

actually consists of four separate bridges: the Knight's Key Bridge proper, a 6,803-foot-long, steel-plate girder span; the 5,935-foot-long Pigeon Key Bridge, which also used a steel-plate girder design; the 13,947-foot-long Moser Channel Bridge, which had a movable truss bridge section (since demolished); and the 9,035-foot-long Pacet Channel Viaduct, a reinforced-concrete arch bridge made up of 210 43-foot-long arches. Work on the Seven Mile Bridge began in 1908 and, despite construction problems resulting from the hot, humid climate, was completed in January 1912. Shortly thereafter, trains began serving Key West.

Although Flagler's dream was an engineering success, the Key West Extension never proved to be financially sound. The Florida East Coast Railroad went bankrupt in 1931, and, following the hurricane of 1935, the firm's receivers decided to abandon operations on the damaged line south of Homestead. Soon afterward the state of Florida acquired the right-of-way and the associated bridges and adapted them for highway use as part of U.S. Route 1. In the early 1980s Flagler's original Seven Mile Bridge was closed and some sections were demolished, but most of the 1912 structure still survives adjacent to the new highway bridge. NR.

ST. AUGUSTINE

Located at the site of a 19th-century ferry, the Bridge of Lions replaced an 1895 wooden trestle built to carry trolley cars to South Beach. In the midst of a major development boom in the 1920s, St. Augustine residents voted to replace the old wooden bridge with a large steel and concrete design. Construction began in July 1925, and by February 1927 the bridge was open for traffic. Since its completion, the structure has been a major symbol of civic pride and it continues to carry highway traffic. The Bridge of Lions, which takes its name from two statues placed at the western portal, is a 1,538-foot-long, steel arch girder structure supported on reinforced-concrete piers. Divided into 23 individual arch spans, it includes a double-leaf bascule bridge in the middle that can be raised to allow for passage of ships along the Matanzas River. NR.

■ **Bridge of Lions**
Across the Matanzas River
On King Street
J. E. Greiner Company
1927

Sculpture standing astride the appropriately named Bridge of Lions in St. Augustine.

■ ■ ■ ■ ■ ■ ■   GEORGIA   ■ ■ ■ ■ ■ ■ ■

COLUMBUS

■ **City Mills Dam**
Across the Chattahoochee
River
At 18th Street and First
Avenue
Hardaway Contracting
Company
1908

In 1828 a small grist mill was built on the site of the present City Mills Dam. Significant rebuilding and alteration occurred in the late 19th century, and by the early 1890s it had become one of west-central Georgia's most important flour mills. In 1894 the owners of City Mills allowed the Columbus Railroad Company to build a powerhouse adjacent to the dam, and this became one of the state's earliest hydroelectric generating facilities. In 1906 the railroad company sold its interest in the site to the Columbus Power Company, which, in conjunction with the City Mills Company, replaced the old wooden dam with a solid-masonry, straight-crested gravity structure, approximately 25 feet high and 700 feet long. This local power company later came under the control of the Georgia Power Company, and in 1951 the hydroelectric plant was abandoned by the latter firm. The company rehabilitated some of its equipment in 1980, however, to continue power production.

Above: Downstream side of the City Mills Dam. Right: Coheelee Creek Covered Bridge, the southernmost covered bridge on a highway in the United States.

HILTON

■ **Coheelee Creek Covered Bridge**
Across Coheelee Creek
On Old River Road
J. W. Baughman, builder
1891

Erected by J. W. Baughman under orders from the Early County board of commissioners, the Coheelee Creek Bridge has the distinction of being the southernmost surviving covered bridge on a public highway in the United States. It is a two-span, modified queen-post truss with an overall length, including approach spans, of 140 feet. The structure's wooden members retain their historical integrity, but a concrete pier has been added at midspan to provide extra support. Since 1957 the Peter Early Chapter of the Daughters of the American Revolution has maintained the bridge for the county. NR.

MULBERRY GROVE

The growth of Columbus, Ga., as a major center of the South's late 19th-century textile industry depended largely on water power developed along the Chattahoochee River. In 1894 the first central-station hydroelectric plant in the city also began supplying power for the local street railway company. During the early 20th century, most of the region's hydroelectric power resources came under the control of George J. Baldwin, a Georgia businessman, and the Boston-based engineering firm of Stone and Webster. With Baldwin acting as its southern agent, the company began building the 70-foot-high, straight-crested concrete gravity Goat Rock Dam about 12 miles north of Columbus in 1910. Completed a year later, the dam and its hydroelectric plant were originally designed to provide power only for the Columbus market. But almost immediately the Goat Rock installation became the center of a much larger regional power system that supplied electricity for towns such as Newnan, Ga., more than 80 miles away. The Goat Rock Dam, named for a now submerged boulder once favored by a local farmer's goats, is owned by the Georgia Power Company and serves customers throughout much of the state.

■ **Goat Rock Dam**
Across the Chattahoochee River
5 miles east of State Route 103
Stone and Webster
1911

Left: Aerial view of the Goat Rock Dam looking upstream. Below: Hydroelectric powerhouse adjoining the Goat Rock Dam.

TALLULAH FALLS

■ **Mathis Dam**
Across the Tallulah River
Near U.S. Route 441, 7 miles
west of Tallulah Falls
Charles O. Lenz
1913

As part of its hydroelectric power development at Tallulah Falls, the Georgia Railway and Power Company constructed a large storage reservoir to impound spring floods and allow maximum development of the Tallulah River's power potential. Lake Rabun, located several miles upstream from the system's main power plant, is formed by a 597-foot-long, 113-foot high, reinforced-concrete Ambursen flat-slab buttress dam. Known as the Mathis Dam, this overflow structure consists of 33 buttresses spaced 18 feet apart and a flat upstream face, which varies in thickness from 18 inches at the top of the dam to 39 inches at the deepest sections. After World War I the Georgia Power Company supplemented its generating capacity on the Tallulah River by building two new storage dams upstream from the Mathis Dam. The largest of these is the Burton Dam, completed in 1919, which impounds a reservoir with a surface area of more than 2,500 acres. Although no longer the only major water storage structure on the river, the Mathis Dam continues to play a role in regional hydroelectric power production.

■ **Tallulah Dam**
Across the Tallulah River
Near U.S. Route 441
Charles O. Lenz
1913

Left: Downstream side of the concrete gravity overflow dam at Tallulah Falls. Right: Original highway running across the slightly curved design, c. 1913.

Given the historical attention lavished on the Tennessee Valley Authority, it is often assumed that there was little electric power development in the South before the 1930s. However, several major hydroelectric power plants operated in the region before World War I. Among the most noteworthy of these was the Georgia Railway and Power Company's 1913 Tallulah Falls development. Built to supply Atlanta with power over a 90-mile-long transmission line, the plant's turbines operated under a head, or pressure, of 600 feet. Water from the Tallulah River is diverted into the tunnel-penstock system by a 129-feet-high, concrete curved gravity dam with an overflow spillway. Although the Tallulah Dam stores a small amount of water in its impoundment reservoir, it is primarily a diversion structure. Storage for the hydroelectric power system is provided by the Mathis Dam (1913), approximately seven miles upstream, and several other more recent impoundment reservoirs on the upper Tallulah River. The dam continues to play a significant role in the Georgia Power Company's electric power grid.

■ ■ ■ ■ ■ ■ ■ KENTUCKY ■ ■ ■ ■ ■ ■ ■

FALLS OF ROUGH

In the 1860s Zenas King became involved in patenting and building bowstring arch truss bridges for highway use. His most widely used design was a tubular arch with a square cross section. Although not practical for use on railroads, this lightweight design was well suited for rural highway bridges. Based in Cleveland, the King Iron Bridge and Manufacturing Company became one of America's most important bridge-building companies, and its tubular bowstring design found use throughout the region east of the Rocky Mountains. The Falls of Rough Bridge is 148 feet long and still retains its original bracing between the top chords. As documented by G. D. Rawlings in a recent inventory of Kentucky's historic bridges, it is both the oldest and longest bowstring arch truss surviving in the state.

■ **Falls of Rough Bridge**
Across the Rough River
On State Route 110
King Iron Bridge and
Manufacturing Company
1877

FULLERTON

In 1914 the Chesapeake and Ohio Railroad began work on this two-span, subdivided Warren through truss as part of a larger project to increase access to the Allegheny coal fields from the north. Designed by Gustav Lindenthal (1850–1935), the bridge is a continuous truss across the center pier and has a total length of 1,550 feet. Historian Carl Condit considers the enormous structure "the ultimate expression of mass and power among American truss bridges." The Sciotoville Bridge continues to carry railroad traffic.

■ **Sciotoville Bridge**
Across the Ohio River
Adjacent to U.S. Route 23
Gustav Lindenthal
1917

Sciotoville Bridge, a cantilever truss. Note how the structure is deeper over the center pier to resist any bending stresses.

Original High Bridge, c. 1905, shortly before it was re-placed. Roebling's suspension bridge towers are visible on the left.

HIGH BRIDGE

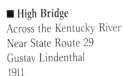

■ High Bridge
Across the Kentucky River
Near State Route 29
Gustav Lindenthal
1911

Flowing northward through the center of Kentucky on its way to the Ohio River, the Kentucky River travels through an imposing gorge more than 200 feet deep. Because the river formed a significant barrier to transportation, as early as the 1850s the Cincinnati Southern Railroad sought to build a permanent bridge at a site about 30 miles southwest of Lexington. At that time John A. Roebling (1806–69) began building a suspension bridge there but, with the coming of the Civil War, all work on the project ceased. It was not until the mid-1870s that the railroad successfully erected a 1,125-foot-long cantilever bridge at the location near Wilmore that came to be known as High Bridge. Designed by Louis F. G. Bouscaren, an assistant engineer on the Eads Bridge (1874) in St. Louis, and C. Shaler Smith, this structure was the first major cantilever bridge in the United States. The bridge was used for 35 years, but in 1911 the Cincinnati Southern completely replaced it with a stronger, more massive design capable of carrying heavier locomotives and rolling stock. None of the iron from the original bridge was used in Gustav Lindenthal's new steel design, although the general configuration of the new structure is similar to the 1875 bridge. The only major change associated with the new span involved raising the height of the trackbed 31 feet to improve the track's gradient. The bridge remains in use by the Cincinnati Southern and Southern railroads. ASCE.

LAKE CITY

The last dam on the Tennessee River before it enters the Mississippi River near Paducah, the Kentucky Dam is a 206-foot-high, straight-crested concrete gravity dam that, in conjunction with earthen dikes, stretches for more than 8,000 feet across the river valley. Like most other dams built by the Tennessee Valley Authority on the main stem of the Tennessee River, it is designed to provide power, flood control and navigation. Kentucky Lake also supplies important recreation opportunities for visitors to the region and is a favorite tourist attraction in the western part of the state.

■ **Kentucky Dam**
Across the Tennessee River
On U.S. Route 62
Tennessee Valley Authority
1944

LOUISVILLE

In 1895 the Louisville and Jeffersonville Bridge and Railroad Company completed construction of a major bridge across the Ohio River for use by the Cleveland, Cincinnati, Chicago and St. Louis Railway. Commonly called the Big Four Railroad, this company used the structure to provide access to southern markets. The main part of the bridge consisted of six pin-connected Pennsylvania through trusses varying in length from 208 to 547 feet. It provided adequate service for almost 30 years, but in the 1920s the bridge company elected to rebuild completely the superstructure to accommodate heavier locomotives and rolling stock.

The new bridge was built on the original structure's piers; thus, the length of the new trusses was the same as those erected in the 1890s. The six main trusses for the new bridge are riveted Pennsyslvania through trusses. Erection of the three largest trusses did not require the building of temporary wooden falsework to support the bridge members during construction; instead, the old trusses were used to stabilize the new superstructure, and the new trusses were placed inside the old ones at a slightly higher elevation. Today, the bridge is no longer used to carry railroad traffic, and local officials are grappling with the problem of what to do with it.

■ **Big Four Railroad Bridge**
Across the Ohio River
On the former Cleveland, Cincinnati, Chicago and St. Louis Railway line, adjacent to U.S. Route 31
L. H. Schaeperklaus and A. M. Westenhoff
1929

Big Four Railroad Bridge resting on 19th-century masonry piers.

■ ■ ■ ■ ■ ■ ■ LOUISIANA ■ ■ ■ ■ ■ ■ ■

NEW ORLEANS

■ Huey P. Long Bridge
Across the Mississippi River
On U.S. Route 90
Modjeski, Masters and Chase
1936

Gov. Huey P. Long has become a renowned political figure in modern American history because of the political power he wielded in Louisiana during the 1930s. At his behest, in 1932 the state legislature authorized the Public Belt Railroad Commission to finance construction of a permanent railroad and highway bridge across the Mississippi River at New Orleans. Because of the state's flat topography, Louisiana contains few large high-level bridges. The enormous stream flow of the river (in excess of a million cubic feet per second) required a concrete caisson foundation extending more than 170 feet below water level. The bridge's steel superstructure is longer than 22,000 feet, and the main channel crossing consists of a 1,850-foot-long cantilever through truss. Because navigation along the river is so important, the center span of the cantilever truss provides 790 feet of horizontal clearance between the piers. The bridge carries both highway and railroad traffic and, after 50 years, still stands as one of the most prominent public works structures in the state. It survived a collision with a large ship and reopened in 1979 after the damaged span was repaired.

Left: Railroad deck of the Huey P. Long Bridge. Right: Train approaching the bridge.

■ ■ ■ ■ ■ ■ ■ MISSISSIPPI ■ ■ ■ ■ ■ ■ ■

BYRAM

■ Byram Bridge
Across the Pearl River
On the Old Byram and
Florence Road
c. 1905

The Byram Bridge is a 360-foot-long steel suspension bridge with a clear span of 200 feet between towers. The design of its concrete-filled cable anchors is based on patents obtained in 1903 by the engineering firm of Schuster and Jacob, located in the nearby town of Fayette. This patent called for the steel cables to be anchored into concrete "deadmen" at both ends of the bridge. Small-scale highway bridges of this type were built throughout Mississippi during the late 19th and early 20th centuries, but only four similar so-called swinging bridges survive in the state. According to material gathered by Jack Gold, former historian for the Mississippi Department of

Swinging bridge at Byram. Its popular name refers to the flexible wire cables.

Archives and History, the Byram Bridge was erected by citizens from Hinds and Rankin counties and the steel hanger rods were forged by De Witt Mason, a local blacksmith. The bridge still carries local traffic and serves as a symbol of a time when bridge construction was a more community-oriented, less bureaucratic exercise in civic improvement. NR.

CHARLESTON

In the early 20th century the Lamb-Fish Lumber Company constructed a large hardwood lumber mill in Charleston, a modest-size city in northwestern Mississippi. As part of a plan to connect the new mill with the Yazoo and Mississippi Valley Railroad, the lumber company commissioned the American Bridge Company to design and build a bridge across the Tallahatchie River about 10 miles east of town. Because the river was used as a navigable waterway, the bridge included a vertical-lift span that could be raised to make way for passing boats. The lumber mill prospered for many years, but following a fire in 1932 it ceased operation. The bridge remained unused until the 1950s, when it was converted to carry highway traffic. The overall bridge is 186 feet long, including a 77-foot-long lift span in the center of the structure. Today, it stands as a symbol of Charleston's once-thriving lumber industry and a rare example of lift-bridge technology in the state. NR.

■ **Lamb-Fish Bridge**
Across the Tallahatchie River
On Paducah Wells Road
American Bridge Company
c. 1905

Left: Lamb-Fish Bridge's vertical-lift span. Right: Road deck passing under the concrete counterweights.

Primeval environment surrounding the Old Hill Place Bridge.

FAYETTE

■ **Old Hill Place Bridge**
Across the South Fork of
Coles Creek
On Hill Road, 1 mile east of
State Route 553 (Natchez
Trace Parkway)
c. 1920

Known locally as a swinging bridge, the Old Hill Place Bridge is a 250-foot-long, clear-span steel suspension structure built to carry rural traffic. The design is based on a 1903 patent taken out by the Fayette-based engineering firm of Schuster and Jacob. Fayette was a center of the state's suspension bridge–building tradition, and at least three contractors in the city — Robert Taylor, J. K. Gallbreath and W. H. Groome and Son — erected bridges of this type in the region. Presumably one of these contractors built the Old Hill Place Bridge, but county records that could document this do not survive. At present, the structure remains in service and is capable of carrying light vehicular loads. NR.

VICKSBURG

■ **Confederate Avenue Steel Arch Bridge**
Across Jackson Road
In Vicksburg National
Military Park
Virginia Bridge and Iron
Company
1903

Fought during the early summer of 1863, the Battle of Vicksburg proved to be a critical turning point of the Civil War. Under the command of Gen. Ulysses S. Grant, Union forces finally took the city after a lengthy siege and thus attained complete control of the Mississippi River. Much of the battlefield site was subsequently preserved as a war memorial and park, and measures were taken to exclude residential and commercial development. However, some thoroughfares were built through the battlefield and, as part of this work, in 1903 the city built the Confederate Avenue Bridge. It is an attractive three-hinged, steel deck arch with a total length of 270 feet. Its main structural members are formed of riveted channel beams similar to those used for many truss bridges. The bridge survives as the oldest known steel arch span in the southeastern United States. Although it no longer carries highway traffic, the structure is maintained for pedestrian use as part of the National Park Service's Vicksburg National Military Park. NR.

Fabricated by the Keystone Bridge Company, the Fairground Street Bridge is one of the oldest metal truss bridges in the South. A two-span, pin-connected Pratt through truss with an overall length of 175 feet, the bridge uses compression members consisting of special Keystone columns similar to the circular cross-sectional Phoenix columns developed by the Phoenix Bridge Company. Keystone columns, however, provide a small airspace between the segments of the riveted compression members. Because of this spacing the Keystone Bridge Company believed it possible to examine the inside of its columns visually and spot easily any deterioration that might threaten the members' structural integrity. The two firms engaged in a patent dispute over the general similarity of their designs, and the Phoenix Bridge Company ultimately proved successful in limiting the widespread use of Keystone columns.

As a result, the Fairground Street Bridge is among only a few surviving structures in the United States that make use of this technology. In a recent study, Jack Elliott of the Mississippi Department of Archives and History documented that the span over Vicksburg's railroad yard dates to at least 1895, which makes it the oldest surviving bridge in the state. However, Keystone columns were most actively promoted by the company during the 1860s and 1870s. This means that the Fairground Street Bridge might easily have been fabricated at an earlier date, erected somewhere else and then moved to its present site in 1895. The bridge still carries some light vehicular traffic, although several of its lower-chord members are deteriorating from lack of maintenance. NR.

■ Fairground Street Bridge
Across the Illinois Central
Gulf Railroad Yard
On Fairground Street
Keystone Bridge Company
c. 1895

WEST POINT

The Tombigbee River is a large waterway that drains much of northeastern Mississippi. During the 19th century it served as an important transportation route for hauling cotton down to the port of Mobile, Ala. Although the coming of railroads to the region in the late 19th century reduced the need for waterborne commerce, the river officially retained its standing as a navigable waterway for many years. Because of this, railroads wishing to cross the river were required to build movable bridges to allow the passage of steamboats. Consequently, when the Columbus and Greenville Railroad extended service across the Tombigbee River near West Point in 1914, it built a large riveted swing bridge with a total length of 218 feet. This movable truss bridge remained in use until the early 1980s when it was abandoned because of the U.S. Army Corps of Engineers' construction of the Tennessee-Tombigbee Waterway. This huge waterway generally follows the route of the Tombigbee River and provides a direct connection between the Tennessee River and the Gulf of Mexico. The return of waterborne navigation as a major form of transportation in the Tombigbee Valley precipitated abandonment of the Columbus and Greenville Railroad

■ Waverly Bridge
Across the Tombigbee River
In Waverly Park, along the
former Columbus and
Greenville Railroad
right-of-way
Wisconsin Bridge and Iron
Company
1914

right-of-way. However, it did not result in the destruction or inundation of the Waverly Bridge. The span now serves as a reminder of earlier transportation systems in the region and is preserved by the Corps of Engineers for pedestrian use as part of Waverly Park. NR.

■ ■ ■ ■ ■ ■   NORTH CAROLINA   ■ ■ ■ ■ ■ ■

CLAREMONT

■ **Bunker Hill
Covered Bridge**
Across Lyle's Creek
Near State Route 1716
Andy Ramsour, builder
1895

Built under authorization from the Catawba County commissioners, the Bunker Hill Covered Bridge is a rare example of the Haupt truss, invented by Hermann Haupt in 1839. Haupt is best remembered for his work in building the Hoosac Tunnel in western Massachusetts in the 1860s and his service as a railroad engineer for the Union army in the Civil War. In addition to this he also developed a truss design that somewhat resembles a Town lattice truss with an auxiliary arch. For unknown reasons, Andy Ramsour resurrected Haupt's design for the Bunker Hill Covered Bridge in the mid-1890s. One of North Carolina's few surviving covered bridges, it no longer carries traffic and is preserved by the county as a local tourist attraction. NR.

Above: Portal of the Bunker Hill Covered Bridge. Right: Aerial view of the Fontana Dam.

FONTANA VILLAGE

■ **Fontana Dam**
Across the Little Tennessee River
Near State Route 28
Tennessee Valley Authority
1945

Not surprisingly, most people associate the Tennessee Valley Authority with Tennessee. A considerable portion of the upper Tennessee River watershed controlled by the TVA, however, is located within North Carolina. The 480-foot-tall, 2,385-foot-long Fontana Dam is the tallest dam in the TVA system and one of the tallest dams in eastern North America. As with almost all dams built by the TVA, it is a straight-crested, concrete gravity structure. Although its massive design is relatively unremarkable when considered within a strictly technological context, the Fontana Dam occupies a particularly beautiful spot in the Smoky Mountains. In fact, historian Carl Condit considers the dam "a perfect symbol of man and nature in harmony." The associated hydroelectric power plant has a capacity of more than 200,000 kilowatts.

Hiwassee Dam, with its sleek, Moderne-style gantry crane, which services the outdoor hydroelectric power plant.

HIWASSEE

Located just east of the Tennessee–North Carolina border, the Hiwassee Dam impounds the headwaters of an important tributary of the Tennessee River. It is a 307-foot-high, 1,287-foot-long, straight-crested concrete gravity dam with a generating capacity of more than 110,000 kilowatts. Although the structure's gravity design is rather ordinary from an engineering point of view, the Moderne facade developed by Roland Wank and the TVA architectural staff is quite stunning. By applying a smooth, starkly handsome veneer to the dam's downstream side, Wank created a design that makes the bulky, massive proportions of the structure appear both functional and artistic at the same time. The Hiwassee Dam continues to serve as a major component of the TVA system.

■ **Hiwassee Dam**
Across the Hiwassee River
Near State Route 294
Tennessee Valley Authority
1940

WATERVILLE

The Walters Dam is a 180-foot-high, 870-foot-long concrete arch dam with a maximum base width of 40½ feet. Built by the Carolina Power and Light Company as part of a hydroelectric power project, the dam supplies water for a power plant located 12 miles downstream on the Pigeon River. This hydroelectric generating system is particularly noteworhty because it operates under a head of 861 feet — one of the highest heads developed east of the Rocky Mountains. Financing and engineering for the project came from the Electric Bond and Share Company, a New York–based utility holding company established by famed financier S. Z. Mitchell (1862–1944). Formally established in 1905, this company became heavily involved in financing electric power systems throughout the United States, and by the mid-1920s it controlled 10 percent of America's electric power industry. The engineering for the Walters Dam is a good example of how EBASCO helped subsidiaries such as Carolina Power and Light develop new power systems. The dam remains an important part of the region's privately owned electric power system.

■ **Walters Dam**
Across the Pigeon River
On State Route 1332
Electric Bond and Share Company
1930

■ ■ ■ ■ ■ ■   SOUTH CAROLINA   ■ ■ ■ ■ ■ ■

ABBEVILLE

■ **Secession Lake Dam**
Across the Rocky River
On County Road 72
James R. Pennell
1940

In early 1933 the Abbeville Power Company received a state charter for a hydroelectric dam and power plant to serve customers in rural western South Carolina. Despite objections from the already existing Duke Power Company and the Southern Public Utilities Company, the Abbeville Power Company, under the guidance of engineer James R. Pennell, soon began building an 80-foot-high, 400-foot-long, reinforced-concrete multiple-arch dam. Construction advanced at a reasonable pace, but in May 1935, with the dam 80 percent completed, the company ran out of money and was forced to abandon the project. The nearby city of Abbeville assumed control of the partially completed dam in hopes of incorporating it into a publicly owned power system. Using funds from the federal Public Works Administration, the city resumed work on the dam and hydroelectric plant in the summer of 1939. Completed in late 1940, the dam soon began generating power for Abbeville and rural customers served by the Little River Electric Cooperative. One of only a few multiple-arch dams built in the eastern United States, the Secession Lake Dam and its power plant are still in active use.

CHARLESTON

■ **Old Cooper River (Grace Memorial) Bridge**
Across the Cooper River
On U.S. Route 17 (Ocean Highway)
Waddell and Hardesty
1929

Often referred to as Old Rollercoaster because of its serpentine appearance, the Old Cooper River Bridge has a total length of more than 10,000 feet. Connecting Charleston with the north side of the Cooper River, the bridge passes over a river channel known as Town Creek, the southern section of Drum Island and the river's main channel. The structure consists primarily of two steel cantilever through trusses with clear spans of more than 760 feet supplemented by lengthy trestle and deck truss approach spans. In 1966 a similar new bridge was built a short distance downstream from the original span. Today, both of these structures are critical components of Route 17, which connects the southeastern United States with points north. The Old Cooper River Bridge recently underwent rehabilitation of its deck and structural system to ensure its long-term preservation and use.

LEXINGTON

■ **Saluda Dam**
Across the Saluda River
On State Route 6
Murray and Flood
1930

In 1927 the Lexington Water Power Company, a subsidiary of the New York–based General Gas and Electric Corporation, began building a huge earthfill dam on the Saluda River about 10 miles west of the state capital at Columbia. Constructed to provide at least 130,000 kilowatts of hydroelectric power, the Saluda Dam is more than 8,000 feet long with a maximum height of 208 feet. At the deepest section the earthen structure is more than one-quarter mile thick. Built in a relatively flat part of central South Carolina, the dam forms a reservoir 41

Left: Aerial view highlighting the buttress design of the Secession Lake Dam. Below left: Original Cooper River Bridge, c. 1930, when it stood alone. Bottom: Distant aerial view of the earthfill Saluda Dam forming Lake Murray in South Carolina, c. 1930.

miles long with a maximum width of 14 miles. Known as Lake Murray, the reservoir covers an area of 78 square miles and contains approximately 750 billion gallons of water (more than two million acre-feet). When it first began operating, the reservoir was the largest artificial body of freshwater in the eastern United States. Today, the Saluda Dam is owned and operated by the South Carolina Electric and Gas Company as part of a regional electric power system.

PINOPOLOS

■ **Pinopolos Dam**
Across the Cooper River
5 miles upstream from U.S.
Route 52
Harza Engineering Company
1941

Located in the coastal lowlands of South Carolina, the Pinopolos Dam is a 140-foot-high, 11,600-foot-long, earthfill gravity structure that creates a reservoir covering almost 100 square miles. Owned by the South Carolina Public Service Authority, it serves a fourfold function: generating hydroelectric power, providing flood control, facilitating navigation and supplementing the local water supply. When completed in 1941, it was among the largest earthfill dams in the Southeast. Since then its designer, the Harza Engineering Company, has developed into a prominent builder of earthfill dams.

TIGERVILLE

■ **Poinsett Bridge**
Across Gap Creek
On County Road 42,
2 miles northwest of
State Route 11
1820

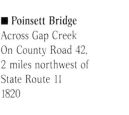

Left: Pointed Gothic arch of the Poinsett Bridge. Right: Rural setting of the small span with its marker.

In 1817 the South Carolina Board of Public Works began building a highway from Charleston through Columbia and north to central North Carolina. Planning for the project was handled by Joel Poinsett, later a U.S. senator and secretary of war, in whose honor the bridge across Gap Creek is named. The picturesque 20-foot-long stone arch structure has the date 1820 inscribed in its keystone. Because the noted architect Robert Mills (1781–1855), initial designer of the Washington Monument, was serving as state architect and engineer for the board of public works at the time, it is possible that he designed it. However, this speculation cannot be confirmed. Today, the Poinsett Bridge is preserved in a local wildlife refuge, and it stands as the oldest known surviving bridge in the southeastern United States. NR.

■ ■ ■ ■ ■ ■ TENNESSEE ■ ■ ■ ■ ■ ■ ■

CALDERWOOD

Designed by the Aluminum Company of America for the regional aluminum smelting operations of Tapoco, Inc., the Calderwood Dam is a 232-foot-high concrete arch dam with a maximum base thickness of approximately 100 feet. The structure is significant because it demonstrated that foundation conditions in the area were generally suitable for building structural arch dams. The Calderwood Dam is located only about 12 miles downstream from the Tennessee Valley Authority's Fontana Dam (1945) in North Carolina, a massive gravity structure. It stands today as one of the tallest arch dams in the eastern United States.

■ **Calderwood Dam**
Across the Little Tennessee River
Adjacent to U.S. Route 129
A. V. Karpov
1930

Downstream side of the Calderwood Dam.

Six-span Walnut Street Bridge resting on masonry piers that carry the structure high above the Tennessee River.

CHATTANOOGA

In 1864 Union forces built a wooden highway bridge across the Tennessee River. Soon after the Civil War it was destroyed by a flood, and for more than two decades no highway bridge spanned the river. A few railroad bridges were built, but otherwise the Tennessee River posed a formidable barrier to overland transportation. In the early 1890s Hamilton County acted to correct this deficiency and authorized construction of a high level bridge across Tennessee's largest stream. Designed by Edwin Thacher (1840–1900), former chief bridge engineer for the Keystone Bridge Company, the Walnut Street Bridge is a six-span, pin-connected, subdivided camelback through truss. On the northern end of the bridge is a trestle approach span 780 feet long that carries the deck over the floodplain. The masonry piers supporting the

■ **Walnut Street Bridge**
Across the Tennessee River
On Walnut Street
Edwin Thacher
1891

trusses were built almost 100 feet above the normal water level in order to reach the top of the bluffs on the Chattanooga side of the river. These high piers also ensured that spring floods would not wash away the superstructure. The Walnut Street Bridge carried highway traffic for more than 75 years, but in 1978 it was removed from use. It is the oldest surviving truss bridge of its size in the South originally built as a highway structure, and efforts are under way to preserve it for pedestrian use.

GOODLETTSVILLE

■ **Mansker Creek Bridge**
Across Mansker Creek
On County Road E 224
c. 1838

The Mansker Creek Bridge is a two-span, masonry arch structure built as part of the Louisville and Nashville Turnpike, also known as the Old Louisville Highway. According to material compiled by Martha Carver, historian with the Tennessee Department of Transportation, "Louisville and Nashville Turnpike" was a term commonly applied to a variety of roads chartered under such names as the Maysville and Washington Turnpike Road Company, the Louisville, West Point and Elizabeth Turnpike Road Company, the Louisville Turnpike Road Company and the Louisville and Elizabeth Company. But despite variations in nomenclature, it appears certain that by the late 1830s highway traffic was being carried by a permanent stone arch bridge over Mansker Creek about 12 miles north of Nashville. The overall bridge is 89 feet long, with each arch having a clear span of 24 feet. The structure, like two others that survive in middle Tennessee, consists of dry-laid stones (that is, no mortar helps bind the masonry together). In recent years the bridge has attracted the care and attention of the Goodlettsville public works department, and city officials are dedicated to preserving it for the enjoyment of future generations. The bridge remains open for light vehicular traffic.

MEMPHIS

■ **Memphis Bridge**
Across the Mississippi River
Adjacent to Interstate 55
George S. Morison
1892

In its northern reaches the Mississippi River is a large yet relatively placid body of water, but after joining with the Missouri River, just above St. Louis, and the Ohio River, at Cairo, Ill., it becomes a powerful, silt-laden waterway that dwarfs all other rivers in North America. The Eads Bridge at St. Louis was completed in 1874, but before the early 1890s no permanent structure crossed the lower Mississippi River. This situation changed with the construction of a huge steel cantilever truss by the Kansas City, Fort Scott and Memphis Railroad System at Memphis. Designed by George S. Morison (1842–1903), this bridge featured a 790-foot-long span near the Tennessee shoreline and two 660-foot-long spans in the center and western sections of the main structure.

Planning for the bridge began in 1885 when Congress issued a charter authorizing construction. Subsequent negotiations with Secretary of War William Endicott over the size and location of the navigation channel

played a major role in determining the final structural design. Endicott stipulated a clear span for navigation of 790 feet and also required that the navigation channel run close to the Tennessee shoreline. As a result of these requirements, Morison developed the somewhat awkward asymmetrical shape of the existing cantilever truss. If nothing else, the bridge is an excellent example of how nontechnical factors, such as navigation requirements, can exert a major influence on engineering design.

Construction of the piers started in late 1888, and by spring work on the steel superstructure was under way. Because of the huge amount of steel in the structure, nine steel companies supplied material used in the bridge. Since it opened in 1892, other bridges have spanned the lower Mississippi, but none has exceeded the Memphis Bridge in grandeur or historical significance. The bridge still carries railroad traffic. ASCE.

Left: Memphis Bridge's 790-foot-long span. Right: Bridge's deck underneath the steel truss construction.

NORRIS

Flowing west out of the Smoky Mountains along the Tennessee–North Carolina border, the Tennesssee River drains a large portion of the southeastern United States, including parts of Virginia, North Carolina, Tennessee, Alabama, Mississippi and Kentucky. Initially settled during the late 18th and early 19th centuries, the Tennessee Valley had become an economic backwater by the beginning of the 20th century. Poorly served by inter-regional transportation systems and possessing few large-scale manufacturing facilities, the valley lay at the heart of Appalachia's insular culture.

The onset of the Depression hit the rural region hard and brought widespread economic devastation to valley residents. In response to this misery, newly elected President Franklin Roosevelt proposed in April 1933 "to create a Tennessee Valley Authority — a corporation clothed with the power of government but possessed of the flexibility and initiative of a private enterprise." An immediate purpose of this innovative governmental agency was to make use of the Wilson Dam (1925) at Muscle Shoals, Ala., originally planned for nitrate production during World War I. But Roosevelt envisaged

■ **Norris Dam**
Across the Clinch River
On State Route 116
U.S. Bureau of Reclamation
1936

Above: Water cascading over the Norris Dam spillway. Above right: Dam with the powerhouse on the right. Right: Upstream side of the dam showing the vertical face and spillway openings.

a more far-reaching authority that would guide water resources, agricultural, industrial and power development throughout the Tennessee Valley. In May 1933 Congress approved plans for the Tennessee Valley Authority in order to promote "the economic and social well-being of the people."

Following its establishment, the Tennessee Valley Authority assumed control of several existing dams, but the first structure actually built by the TVA was the Norris Dam in northeastern Tennessee. Designed to provide power, flood control and recreation, it is a 1,872-foot-long, 265-foot-high, straight-crested concrete gravity dam with an overflow spillway in the middle of the structure. Because of the Roosevelt administration's desire that construction begin as soon as possible, structural design work for the Norris Dam was handled by the U.S. Department of the Interior's Bureau of Reclamation. The modern architectural treatment of the design, however, came from the TVA's own architectural staff under the direction of Roland Wank. Originally called the Cove Creek Dam, the Norris Dam does not refer to any resident or locality in the Tennessee Valley. Instead, it is named after Sen. George W. Norris of Nebraska, a prominent supporter of federal public works development. The dam generated power for distribution to farmers in the region through rural electrification projects, and it became an important agent of change and progress. However, the dam's reservoir also flooded out numerous established farms in the upper Clinch River Valley and caused the dislocation of many long-time residents. Thus, construction of the Norris Dam is a classic example of how large-scale dam projects can simultaneously spawn both benefits and hardships for a region. It remains today in active service. ASCE.

ONEIDA

The Cincinnati Southern Railway opened for service between Cincinnati and Chattanooga, Tenn., in 1880 and subsequently prompted the construction of numerous tributary railroads in eastern Tennessee. Among these was the Oneida and Western Railroad, an offshoot of the Tennessee Stave and Lumber Company, built to facilitate logging operations in the relatively inaccessible region west of Oneida. Construction of the line began in late 1913, and by mid-1915 the company completed the crossing over the Big South Fork River. This structure consists of a 201-foot-long, pin-connected, double-intersection Pratt through truss and a few short approach spans. The wrought-iron truss members and the basic structural design indicate that the Big South Fork Bridge was built sometime in the 1880s and moved to the Oneida and Western right-of-way in 1915. The railroad line was abandoned in 1954, and the bridge then became part of the Scott County highway system. It is now being preserved as a historic structure within the newly formed Big South Fork National River and Recreation Area.

■ **Oneida and Western Railroad Big South Fork Bridge**
Across the Big South Fork River
In the Big South Fork National River and Recreation Area, east of U.S. Route 27
Nashville Bridge Company
c. 1885; reerected 1915

PARKSVILLE

Even before World War I, some privately financed utility companies made attempts to develop the hydroelectric power potential of streams flowing west out of the Smoky Mountains. An example of this is the Ocoee Dam No. 1, a 135-foot-high, 840-foot-long concrete gravity dam built by the Eastern Tennessee Power Company. The main portion of the dam is straight crested, but the overflow spillway section is built as a curved gravity dam. Shortly after completion of the hydroelectric power plant at Ocoee No. 1, the power company built another plant, Ocoee No. 2, approximately 10 miles upstream. In the 1930s both facilities were taken over by the Tennessee Valley Authority and have remained under the agency's control ever since. Compared with other TVA plants, Ocoee No. 1 is relatively small. It makes an important contribution to the system's generating capacity, however, and serves as a useful reminder that the TVA did not invent electric power generation in the Tennessee Valley watershed.

■ **Ocoee Dam No. 1**
Across the Ocoee River
Adjacent to U.S. Route 64
J. G. White Engineering Corporation
1911

Ocoee Dam No. 1, with its powerhouse on the left and curved spillway on the right.

■ ■ ■ ■ ■ ■ ■   VIRGINIA   ■ ■ ■ ■ ■ ■ ■

CARTERSVILLE

■ Cartersville Bridge
Across the James River
Adjacent to State Route 45
1884

The Pratt truss is most commonly thought of as an all-metal design, but when Thomas and Caleb Pratt prepared their patent application in 1844, they originally stipulated the use of wooden vertical compression members and wrought-iron diagonal tension members. The dominant design in the field of combination wood and iron trusses was the Howe truss, and the all-metal Pratt truss did not become common until later in the 19th century. Some wooden Pratt trusses were built, however, and the Cartersville Bridge is a good example of this type of structure. Constructed at the site of a former ferry, the 843-foot-long, combination wood and iron bridge carried highway traffic until June 1972, when Hurricane Agnes carried away the center four of its six spans. The two remaining end spans, each about 137 feet in length, are now preserved as a local historic site. Although much of its grandeur is lost, the Cartersville Bridge still stands as an example of how Thomas and Caleb Pratt originally visualized their truss. NR.

CHESAPEAKE BEACH

■ Chesapeake Bay Bridge and Tunnel
Across and under the
Chesapeake Bay
On U.S. Route 17
Sverdrup and Parcel
1964

Stretching 17½ miles across the mouth of the Chesapeake Bay in eastern Virginia, this monumental achievement in civil engineering includes more than 13 miles of concrete trestles and steel trusses, two tunnels each more than a mile in length, four man-made islands and a mile-and-one-half-long earthfill causeway. Built along the route of the old Kiptopeke–Little Creek Ferry, the structure provides a permanent transportation link between Cape Henry, to the east of Norfolk, and Cape Charles, at the southern tip of Virginia's Eastern Shore. Project planning began in 1956, and the bridge-tunnel was opened for traffic eight years later. Tunnels, instead of bridges, for portions of its length ensure that storms or sabotage can never block access to the important naval base at Newport News. Operated and maintained by the Chesapeake Bridge-Tunnel Commission, the facility remains in service as one of the largest and most impressive transportation facilities in the world.

COVINGTON

■ Old Humpback Bridge
Across Dunlap Creek
Adjacent to U.S. Route 60
1857

Built as part of the James River and Kanawha Turnpike that connected piedmont Virginia with the Ohio River Valley, the Old Humpback Bridge is the oldest surviving covered bridge in Virginia. For many years, it was believed to date from the mid-1830s, but recent research indicates the bridge was built in 1857. Before the Civil War, it was only one of many covered bridges in the region. Both the North and South destroyed numerous wooden bridges during the conflict, however, and by the early 20th century the Old Humpback Bridge existed as a rare example of a surviving antebellum covered bridge in

Top: Cartersville Bridge before 1972. Above and left: Well-preserved Humpback Bridge in southwestern Virginia, with its masonry foundations and curved multiple king-post design.

Virginia. Now maintained as part of a local park, it is a 100-foot-long, multiple king-post truss. Its unusual name derives from the distinctive curve built into the structure's top and bottom chords. Presumably to help prevent the bridge from sagging under heavy wagon loads, the center is eight feet higher than the two ends of the structure. Two other humpback bridges were built on the James River and Kanawha River Turnpike, but they are now gone, making the Old Humpback Bridge the only surviving example of this type of design in the United States. NR.

EAGLE ROCK

■ Phoenix Bridge
Across Craig Creek
⅛ mile north of State
Route 615
Phoenix Bridge Company
1887

Originally built as part of a minor Chesapeake and Ohio Railroad spur line, this single-span, pin-connected Pratt through truss is approximately 150 feet long. It is an excellent example of a truss erected using Phoenix columns for the compression members. Made of wrought-iron segments rolled along a circular arc and riveted together, this type of column was strong but did not connect easily with other structural members and was eventually abandoned. The Phoenix Bridge in Eagle Rock is now maintained as part of a local highway operating along the old railroad right-of-way. NR.

Right: Phoenix Bridge, with its ornate portal design. Below: Connection between the lower chord and the hipped vertical.

Riveted Warren truss approach span of the bridge.

NOKESVILLE

As railroad locomotives and rolling stock grew heavier in the late 19th and early 20th centuries, it was not uncommon for older railroad truss bridges to be adapted for lighter highway traffic. For example, the Nokesville Bridge was built by one of America's largest 19th-century bridge companies and is an excellent example of a small-scale, pin-connected Pratt through truss. Only 74 feet long, it is believed to have been used originally to carry railroad traffic. Moved to its present site in the early 20th century, it now carries highway traffic across the Southern Railroad tracks about 35 miles southwest of Washington, D.C. NR.

■ **Nokesville Pratt Truss Bridge**
Across the Southern Railroad line
On State Route 646
Keystone Bridge Company
1882

Nokesville's bridge and a detail of its abutment design.

PEDLAR MILLS

Although not large compared with other cities in the Northeast, Lynchburg built one of America's first municipal water supply systems in the 1830s. Construction of the Pedlar Dam in the early 20th century represents a continuation of Lynchburg's civic efforts to provide a safe and secure supply of water. When originally built, the dam had a maximum height of 61 feet from deepest foundations to the top of the spillway. In the mid-1960s it was raised to its present height of 85 feet. It is a straight-crested, gravity structure built using concrete imbedded with large stones, a structural material often called cyclopean concrete. Water from the dam flows through a 12-mile-long aqueduct before it reaches the city proper. By building the dam in a sparsely populated area, it was possible to help ensure an uncontaminated municipal water supply.

■ **Pedlar Dam**
Across the Pedlar River
½ mile north of State Route 607
H. L. Shaner
1904, c. 1965

Pedlar Dam before it was raised, c. 1910.

MIDWEST

One of the three main arches of the Eads Bridge (1874) in St. Louis. This monumental structure was the first permanent crossing of the Mississippi River below its confluence with the Missouri River.

■ ■ ■ ■ ■ ■ ■   ILLINOIS   ■ ■ ■ ■ ■ ■ ■

CARLYLE

**■ General Dean
Suspension Bridge**
Across the Kaskaskia River
Adjacent to U.S. Route 50
D. Griffith Smith, builder
1859

Constructed as part of the old Vincennes to St. Louis trail, a highway route now followed by U.S. Route 50, the General Dean Suspension Bridge is a 264-foot-long iron suspension bridge supported on stone and brick towers. It served the traveling public for more than 60 years until replaced in the 1920s by a new bridge located slightly downstream. In the late 1940s its distinctive appearance and historical significance prompted local citizens to rehabilitate the structure, adding new cables and encasing the towers in concrete. With financial help from the state legislature, this work was accomplished in the late 1950s, and the structure has been maintained as a historic site ever since. At that time it was named in honor of Maj. Gen. William Dean, a prominent resident of Carlyle who served in the Korean War. NR.

Dean Bridge, with its wooden traffic deck.

CHICAGO

■ Cermak Road Bridge
Across the South Branch
of the Chicago River
On Cermak Road
William Scherzer
1906

By most standards the Chicago River is a fairly small stream with a limited watershed. In contrast to most rivers in Illinois, it does not flow south toward the Gulf of Mexico; instead, it empties into Lake Michigan after draining much of the land that makes up Chicago. The river flows around the city's famous Loop district, prompting construction of numerous movable bridges that could accommodate the needs of both vehicular and ship traffic. In the late 19th century Chicago engineers experimented with a variety of movable bridges, including vertical-lift structures and swing spans. By 1900 the city determined that bascule lift bridges were the most efficient, and they became the standard type built in Chicago during the 20th century.

One of the oldest movable bridges in the metropolitan area is the 216-foot-long Cermak Road Bridge. As a

bascule bridge, its superstructure is balanced in such a way that the span can be easily lifted to provide clearance for ships to pass under. First developed by Chicago engineer William Scherzer (1858–93), the Scherzer type of bascule bridge operates by having the span roll back along its foundation support. As a result this type of structure is often called a Scherzer rolling lift bridge. As with many other bascule bridges, the Cermak Road Bridge actually consists of two rolling lift structures, one on each side of the river. When the bridge is closed — that is, ready to carry vehicular traffic — the two sides of the bridge meet in the center of the river and form a continuous span. When the bridge is open, the two sides lift up and provide clearance for passing ships. The Cermak Road Bridge is the oldest surviving Scherzer rolling lift bridge operated by the city.

Left: Cermak Road Bridge across the Chicago River, 1914. Right: Cortland Street Bridge as seen by approaching travelers, 1914.

Another bascule bridge, the 128-foot-long Cortland Street Bridge was built to replace an earlier swing bridge at the site. Among the most important advantages over standard swing-span designs afforded by the new double-leaf bascule design was the elimination of a pivot pier in the middle of the river, which allowed substantially more room for the passage of ships. The two bascule spans making up the bridge (one is located on each side of the river) are moved by machinery located below the traffic deck. Each leaf is capable of being rotated upward 75 degrees by a fixed pinion that meshes with a rack located on the curved heel of each span. Although the machinery in the substructure is still intact, the Cortland Street Bridge is no longer operable, and the two leaves are clamped together in the closed position. According to the Illinois Historic Preservation Agency, it is the oldest bascule bridge in the state. ASCE.

■ **Cortland Street Bridge**
Across the North Branch of the Chicago River
On Cortland Street
Edward Wilmann and John Ericson
1902

CLARK CENTER

Construction of the National Road began in western Maryland in 1811 and gradually continued westward toward the Mississippi River. In the 1830s work began on extending the road through Illinois. As part of this project, the Old Stone Arch Bridge was built near Clark Center in Clark County, which abuts the Illinois-Indiana border. This 18-foot-long stone structure was one of the

■ **Old Stone Arch Bridge**
Across Mill Creek Wash
Adjacent to U.S. Route 40
c. 1830

Left: Stone arch near Clark Center, built as part of the National Road. Right: Swing span of the Rock Island Bridge, c. 1905.

first bridges on the National Road constructed in the state. It remained in use until the mid-20th century, when Route 40 was rebuilt along a new alignment. Although not a large structure, it survives as an important engineering relic dating to an early era in the state's transportation history. NR.

ROCK ISLAND

■ **Rock Island (Government) Bridge**
Across the Mississippi River
In the Rock Island Arsenal,
at Rodman and Fort
Armstrong Avenues
Ralph Modjeski
1896

In 1868–72 the U.S. War Department built a new railroad bridge connecting the Rock Island Arsenal with the city of Davenport, Iowa, on the west bank of the Mississippi River. The structure supported a single railroad track and a vehicular roadway but soon became a major transportation bottleneck for the army, so in 1895 work began on a larger bridge to be built on the stone piers of the 1872 span. Designed by Ralph Modjeski (1861–1940) and fabricated by the Phoenix Bridge Company, the bridge opened for traffic in December 1896. It consists of two riveted Pratt through trusses, five riveted Baltimore through trusses and a pin-connected Baltimore swing truss. All together, the bridge is almost 2,000 feet long. In addition to its double-track railroad, the bridge also supports a vehicular roadway beneath the railroad tracks. The bridge is still owned, operated and maintained by the federal government for transportation between the Rock Island Arsenal and the Davenport Arsenal in Iowa. The bridge remains little changed and survives as one of the oldest bridges connecting Illinois and Iowa.

WATERLOO

■ **Fountain Creek Bridge**
Across Fountain Creek
Adjacent to State Route 156
Herman Gerlab, builder
1849

Located along the bottom lands of the Mississippi River in south-central Illinois, Monroe County contains numerous limestone quarries and outcroppings that yield masonry of excellent quality. As historian Keith Sculle of the Illinois Historic Preservation Agency has documented, the ready availability of limestone led to the construction of approximately 100 masonry highway bridges in the county during the 19th and early 20th centuries. Among the earliest and most impressive of these structures is the 42-foot-long Fountain Creek Bridge, built as part of the old Dennis Hollow Road. The bridge remained in use for more than 70 years until bypassed by construction of a new highway in the 1920s. Today, it stands preserved as an impressive monument to Monroe County's long tradition of masonry bridge building. NR.

■ ■ ■ ■ ■ ■   INDIANA   ■ ■ ■ ■ ■ ■

ANNAPOLIS

By the late 19th century iron and steel bridges had achieved great popularity throughout the United States. But despite the plethora of engineers and bridge companies promoting metal truss bridges, covered wooden bridges still remained popular in many areas, including Parke County in central Indiana near the Illinois border. Almost 60 covered bridges are known to have been built in this rural county during the past 140 years. The Jackson Bridge, a 207-foot-long Burr arch truss, is the oldest covered bridge in Parke County located on its original site and is also the longest single-span wooden bridge in the county. J. J. Daniels built the Jackson Bridge and nine other covered bridges in the county that still survive. In 1977 the county highway department repaired the floor beams and deck structure to accommodate modern traffic loads, but the structure still retains much of its structural integrity. NR.

■ **Jackson Bridge**
Across Sugar Creek
On Parke County Road 83
J. J. Daniels, builder
1861

AURORA

The Laughery Creek Bridge is a rare example of a triple-intersection Pratt truss, a variation on the double-intersection form. To build economical long-span, simply supported truss bridges, it is advantageous to design very tall structures. The amount of material required for a given span length thus can be reduced substantially over a shallower truss of equal length. When using Pratt trusses, engineers quickly realized that single-intersection designs (i.e., those in which the diagonal tension members extend the length of only one panel) could not be used for very tall structures unless the panel length was quite large. Because long panels required lengthy, uneconomical stringers between the trusses' floor beams, means were sought to reduce panel lengths for tall trusses. This led to the development in the late 1840s of the double-intersection Pratt truss by Squire Whipple (1804–88), in which the diagonal tension members extended across two panels. Whipple's innovation essentially reduced the panel lengths by 50 percent and allowed for less-expensive deck systems.

The double-intersection Pratt truss was used extensively in the late 19th century, and, in a few instances, its success prompted construction of triple-intersection Pratt trusses. As the name indicates, this type of bridge

■ **Laughery Creek Bridge**
Across Laughery Creek
Adjacent to State Route 56
Wrought Iron Bridge Company
1878

Elevation drawing of the triple-intersection Pratt through truss across Laughery Creek.

had diagonal members extending across three panels and theoretically facilitated the use of even taller trusses. The triple-intersection Pratt truss never became popular because its extremely long diagonals were not particularly rigid, especially under heavy, fast-moving loads. In addition, the subdivided panels used in the Baltimore and Pennsylvania trusses also allowed for the construction of tall trusses with relatively short panel lengths.

The Laughery Creek Bridge has an overall length of 300 feet and a height of 40 feet; its panels are slightly more than 14 feet long. Funding for the bridge was authorized by the commissioners of Ohio and Dearborn counties to replace an old Howe truss at the site that had collapsed. The bridge functioned adequately for several decades before being replaced by a new highway bridge about one-fourth mile downstream. Today, the Laughery Creek Bridge is closed to traffic, but it survives as an example of a technology that seemed to offer structural advantages at one time but ultimately proved to be unnecessary and unpopular. NR.

CHILE

■ **Paw Paw Creek Bridge**
Across Paw Paw Creek
On 800 North Road,
near Paw Paw Church
Wrought Iron Bridge
Company
1874

Paw Paw Creek Bridge, an excellent example of a bowstring arch truss fabricated by David Hammond's Wrought Iron Bridge Company.

In 1874 the Miami County commissioners contracted with the Wrought Iron Bridge Company for an 111-foot-long bowstring arch truss bridge across Paw Paw Creek near the Paw Paw Methodist Church (1846). Founded by David Hammond (1830–c. 1905) in 1866, this company built a wide variety of metal truss bridges throughout the United States during the late 19th century. The firm actively promoted its own patented type of bowstring design, featuring top-chord compression members riveted into a tubular shape. Hammond guided the firm until the early 1880s, and eventually it was taken over by Andrew Carnegie's American Bridge Company. The Paw Paw Creek Bridge is a good example of the Wrought Iron Bridge Company's own special type of bowstring arch truss bridge. In 1980 the structure was rehabilitated by the county and received a new wooden deck supported on new steel stringers. Although loads are limited on the bridge, it still carries light vehicular traffic. NR.

MECCA

The Mecca Bridge is a 150-foot-long Burr arch truss erected by J. J. Daniels, a prominent local builder. For reasons that remain unclear, the vast majority of Indiana's surviving long-span covered bridges employ the Burr system. These designs have proven to be extremely sturdy structures capable of carrying relatively heavy highway traffic. The Mecca Bridge remained a part of the county highway system for almost 100 years before being retired from use in the mid-1960s. It is still preserved as a pedestrian bridge. NR.

■ **Mecca Bridge**
Across Big Raccoon Creek
On Parke County Road 89
J. J. Daniels, builder
1873

MIDWAY

The Phillips Bridge, only 43 feet long, is a small multiple king-post truss. It cannot be considered a major example of covered-bridge technology but is representative of many small rural wooden bridges built in the 19th and early 20th centuries. It was erected by Joseph A. Britton, a local covered-bridge builder who, in conjunction with his sons, constructed 14 wooden spans in Parke County. The Phillips Bridge still carries a small amount of local traffic. NR.

■ **Phillips Bridge**
Across Big Pond Creek
On Parke County Road 89
Joseph A. Britton, builder
1909

MILROY

Archibald M. Kennedy came to Indiana as a child in the 1820s and eventually settled with his family in Rushville, a small town in the southeastern part of the state. He worked as a carpenter in Rush County and, among other projects, built his own Italianate house in 1864. When he was in his early 50s, he began building wooden covered bridges for local highway use in the region. He erected his first bridge in 1870 in Union County and initiated a Kennedy family bridge-building tradition, which lasted for almost 50 years. During this time Kennedy, his sons Charles and Emmett, and Emmett's sons Karl and Charles were responsible for building at least 58 wooden bridges in the state over a period of 40 years. Of the 13 that survived into the 1980s, six are located in Rush County. The Ferree Covered Bridge, the oldest surviving

■ **Ferree Covered Bridge**
Across the Little Flatrock River
On Base Road
Archibald M. Kennedy, builder
1873

Ferree Covered Bridge, a Burr arch truss whose future is uncertain.

Kennedy-built bridge in Indiana, is 87 feet long and a good example of a Burr arch truss design. At present the bridge remains open to light vehicular traffic, but county commissioners are seriously contemplating replacing the span with a new concrete bridge. While the Rush County Heritage Society is attempting to rally public support for preserving the Ferree Covered Bridge, its future remains clouded. NR.

MOSCOW

■ **Moscow Covered Bridge**
Across the Big Flatrock River
At Rush County Roads 875 and 625
Emmett L. Kennedy, builder
1886

By the mid-1880s the covered-bridge building tradition started by Archibald Kennedy had passed on to his son Emmett. In 1886 Emmett began work on a two-span, Burr arch truss bridge in Moscow, a small town southwest of the Kennedy family home in Rushville. The Moscow Covered Bridge, which is 380 feet long, is the only multispan bridge built by the Kennedys that survives; eight others have been demolished. In addition, it is the longest of the family's surviving bridges. The structure remains open to light vehicular traffic. NR.

ROCKVILLE

■ **Leatherwood Station Bridge**
In Billie Creek Village
Adjacent to U.S. Route 36
Joseph A. Britton
1899

The Leatherwood Station Bridge was built as part of a turnpike between Rockville and Newport, the county seat of nearby Vermillion County. The Parke County commissioners let the contract to Joseph A. Britton (1839–1929) and authorized expenditure of $900 for a 50-foot-long bridge across Leatherwood Creek. Britton was born near Rockville and spent much of his life building covered bridges in the Parke County region. As with most other Britton structures, the Leatherwood Station Bridge is a Burr arch truss. Because of deteriorating abutments, the bridge was closed to traffic in 1979 and moved to nearby Billie Creek Village on the outskirts of Rockville. It is now maintained as part of a museum village that re-creates the ambience of a 19th-century Indiana crossroads community. NR.

Leatherwood Station Bridge in its new location as part of a museum village.

RUSHVILLE

By 1916 the use of wood for new highway bridges was something of an anomaly in the Midwest. Steel and concrete structures were favored by the vast majority of county highway engineers; wooden bridges, with their frequent maintenance problems, held little allure. But in Rush County the Kennedy family tradition of building covered bridges continued and found support among the county commissioners. The Norris Ford Bridge is 154 feet long and employs the Burr arch truss for its structural system. Amazingly, the design is almost identical to the type of covered bridge that Archibald M. Kennedy began building in the early 1870s. On the portal are small Italianate-style brackets similar to those used for Archibald Kennedy's 1864 house and for many other bridges built by the family. The Norris Ford Covered Bridge is still open to light vehicular traffic. NR.

■ **Norris Ford**
Covered Bridge
Across the Big Flatrock River
On Rush County Road 150
Emmett Kennedy, Karl Kennedy and Charles Kennedy, builders
1916

During the first part of the 1870s, Archibald M. Kennedy received a considerable number of contracts from local counties in need of small highway bridges. The aftershock of the financial panic of 1873 left Kennedy, along with many other builders and bridge companies, in a precarious position. Finally, in 1877 Kennedy received authorization from the Rush County commissioners to build the Smith Covered Bridge. Working with his son Emmett, he erected a 124-foot-long Burr arch truss across the Big Flatrock River on the western edge of Rushville. With this contract the Kennedy family began a decade of intense work that resulted in the construction of 40 covered bridges. The Smith Covered Bridge is now closed to traffic, and the county commissioners reportedly have plans to bypass the structure by building a new concrete bridge nearby. NR.

■ **Smith Covered Bridge**
Across the Big Flatrock River
On Rush County Road 300
Archibald M. Kennedy and Emmett Kennedy, builders
1877

Interior of the Smith Covered Bridge, a Burr arch truss.

■ ■ ■ ■ ■ ■ ■ ■ IOWA ■ ■ ■ ■ ■ ■ ■ ■

FREEPORT

■ **Freeport Bowstring Arch Truss Bridge**
Across the Upper Iowa River
¼ mile north of Freeport
Wrought Iron Bridge
Company
1878

The Freeport Bridge is one of several bowstring arch trusses built by the Wrought Iron Bridge Company that, until recently, survived in Winnishiek County. With an overall length of 160 feet, it is the longest of its type in the state. The Freeport Bridge retains its original cross bracing between the upper chords, and the company nameplate is intact. The road deck was rebuilt in 1977, but otherwise the structure retains its original design integrity. In late 1987 the bridge was closed to vehicular traffic, and future plans for the bridge are uncertain. It is hoped that demolition will be averted and the means found to preserve the structure in its original location. NR.

KEOKUK

■ **Keokuk Dam**
Across the Mississippi River
Adjacent to U.S. Route 136
Hugh L. Cooper
1913

During the 19th century steamboat operators on the upper Mississippi River were painfully aware of the Keokuk Rapids located between Keokuk, Iowa, and Hamilton, Ill. In the early 20th century some engineers and businesspeople realized that this obstacle to navigation could constitute the source of a major hydroelectric power project. Built by the Mississippi River Power Company between 1910 and 1913, the Keokuk Dam became the first major dam on the Mississippi River below St. Paul, Minn., and retained this unique status until the 1930s, when the U.S. Army Corps of Engineers built a series of navigation dams along the river. Designed by Hugh L. Cooper, the Keokuk Dam is a 53-foot-high, 4,696-foot-long, concrete, straight-crested gravity overflow structure. The structure helped further Cooper's international reputation in large overflow dam design and presaged later work he undertook for the Soviet

Original hydroelectric
powerhouse, built as part of
the Keokuk Dam.

Union in the early 1930s. Today, the Keokuk Dam is still used for power production and is designated Dam No. 19 by the Corps of Engineers, which also operates navigation locks at the site. NR.

MONTICELLO

Born in Bucks County, Pa., in the 1820s, Reuben Ely, Sr.,
later moved to eastern Iowa and established a family farm
near Monticello. In 1893 he and his son built this three-
span masonry arch bridge across Deer Creek. The
structure is 60 feet long and is distinguished by three
centered elliptical arches. Stone for the piers and
spandrel walls came from the local stream bed, while the
dressed stone used for the voussoirs (i.e., the stones that
form the structural part of the arch) was obtained from
the quarries at Anamosa, about 10 miles away. The Ely
family still keeps a watchful eye on the bridge and
provides minor maintenance when needed. The pier
foundations have recently been strengthened with concrete to protect the bridge from water scouring. It still
carries local highway traffic. NR.

■ **Ely's Stone Bridge**
Across Deer Creek
On Hardscrabble Road
Reuben Ely, Sr., and Reuben
Ely, Jr., builders
1893

Left: Ely's Stone Bridge, with
its new concrete base and
supports. Right: Lincoln
Highway Bridge, with its
distinctive balustrade.

TAMA

In 1913 automobile industry officials and others interested in promoting long-distance automobile touring
organized the Lincoln Highway Association. This nonprofit corporation was dedicated to building a well-
marked, paved highway across the United States from
New York to San Francisco. As part of its efforts to
increase public support for improved highways, the
association helped fund "demonstration sections" of
concrete roadway in various states and regions. It also
worked to publicize highway development in less substantive ways and encouraged cities and small towns to
boost automobile travel as a means of local economic
development.

The Lincoln Highway Bridge at Tama is a result of
such boosterism. To distinguish itself from the hundreds
of other small towns along the highway, Tama built the
20-foot-long, reinforced-concrete girder bridge across
Mud Creek with guardrails that spell out "Lincoln
Highway." The distinctive span carried traffic on the
transcontinental highway for many years until U.S. Route
30 was upgraded along a new alignment. It now serves
local transportation needs in a quiet part of the city. NR.

■ **Lincoln Highway Bridge**
Across Mud Creek
On East Fifth Street
Iowa Highway Department
1915

■ ■ ■ ■ ■ ■   KANSAS   ■ ■ ■ ■ ■ ■

AUBURN

■ **McCauley Bridge**
Across the North Branch
of the Wakarusa River
East of Federal Aid System
Road 514
Topeka Bridge and Iron
Company
1916

Left: Main span of the Austin
Bridge. Right: Two-span,
reinforced-concrete McCauley
Bridge, whose construction
followed David Luten's
patents.

The McCauley Bridge is a 100-foot-long, two-span, reinforced-concrete deck arch built in accordance with patents obtained by Daniel Luten. These patents described concrete arch structures imbedded with steel reinforcing in areas where tensile stresses were likely to occur. In technical terms his designs were not radically different from other reinforced-concrete bridges being built in the early 20th century. Luten, however, was very active in promoting his designs throughout the United States, and he found great success in using a variety of local contractors to construct bridges using his patents. One of these was the Topeka Bridge and Iron Company, which received the commission from Shawnee County for the McCauley Bridge. Named after a nearby land-owner, the bridge remained in use until 1948, when it was bypassed during a highway realignment project. Now located on private property near a local nursing home, the abandoned structure is one of the oldest surviving reinforced-concrete bridges in the state. NR.

CHANUTE

■ **Austin Bridge**
Across the Neosho River
4 miles south of State
Route 39
King Iron Bridge and
Manufacturing Company
1872; reerected 1910

Shortly after the Civil War, large sections of southeastern Kansas were removed from Indian reservations and opened up to Anglo-American settlement. The Neosho River, a tributary of the Arkansas River, drains much of this region and formed a significant barrier to local commerce and transportation. As documented by Robert Hosack, former president of the Neosho Valley Historical Society, in August 1872 Neosho County awarded a contract to the King Iron Bridge and Manufacturing Company for a metal truss bridge across the river at the present site of State Route 39. Although based in Cleveland, the company had established an auxiliary plant in Kansas during the early 1870s—first in Iola, later in Topeka — to serve markets in the expanding frontier. The 160-foot-long bowstring arch truss — among the longest such trusses known to survive in the United States—carried highway traffic in its original location for more than 40 years before being moved in 1910 to its present site about four miles downstream. Today, the bridge, and a 79-foot-long Pratt truss approach span, are preserved as part of a county park and are accessible to pedestrians. NR.

DOUGLASS

Built using locally quarried stone, this 31-foot-long single-span arch bridge is a good example of vernacular engineering in the Midwest. Nothing is known about the builder, but the bridge survives as an attractive structure that still plays a role in the Butler County highway system. Although Kansas is perceived as a flat expanse of prairie land, the state actually has a relatively varied terrain, especially in its eastern sections. This terrain, with its numerous rock outcroppings, fostered the construction of many small-scale stone arch bridges, such as the Polecat Creek Bridge, during the late 19th century. NR.

■ **Polecat Creek Bridge**
Across Polecat Creek
On a local road, 7 miles
north of State Route 15,
at Udall
c. 1900

Above: Graceful masonry
arch across Polecat Creek.

ELGIN

In a recent survey of highway bridges, the Kansas Department of Transportation located 73 examples of reinforced-concrete Marsh rainbow arch bridges in the state built between 1917 and 1940. In 1883 James B. Marsh (1856–1936) began working for the King Iron Bridge and Manufacturing Company and promoted the company's metal truss designs throughout the Mississippi River Valley. In 1896 he started his own bridge-

■ **Cedar Creek Bridge**
Across Cedar Creek
On Federal Aid System
Route 96
Marsh Engineering Company
1927

Rainbow arch across Cedar
Creek.

Cedar Creek Bridge, with its Marsh rainbow arch.

building firm in Des Moines and, as the chief proprietor of the Marsh Engineering Company, received a patent for the Marsh rainbow arch bridge in August 1912. This design resembled a metal bowstring arch truss structure but used reinforced concrete for its main structural members. Technically, Marsh probably should not have been given a patent, because other engineers had previously built structures of similar design. But Marsh actively promoted the technology and became responsible for the vast majority of rainbow arch bridges built in the Midwest and Great Plains.

Several Marsh rainbow arch bridges have since been listed in the National Register of Historic Places in recognition of their role in Kansas's economic development. The 82-foot-long Cedar Creek Bridge is a representative example of the structural form. When an earlier bridge at this crossing washed out in late 1926, the Marsh Engineering Company received the contract for the replacement span in May 1927. After flooding disrupted construction in late August, the rainbow arch structure was finally ready to carry traffic in December. Since then, it has continued to serve the needs of Chautauqua County residents. NR.

FULTON

■ **Long Shoals Bridge**
Across the Little Osage River
6 miles east of U.S. Route 69,
at the site of Long
Shoal Ford
Midland Bridge Company
1902

Located only a few hundred yards west of the Kansas-Missouri state line, the Long Shoals Bridge is a 176-foot-long, pin-connected Parker through truss. Erected by the Midland Bridge Company of nearby Kansas City, Mo., it is the largest surviving pin-connected Parker truss in Kansas. Also of note are the vertical, rather than inclined, end posts for the trusses. By the beginning of the 20th century, the great majority of through trusses were designed with inclined end posts in order to help minimize the amount of steel required for the structure. In this context the Long Shoals Bridge is a bit of an anachronism, reflecting bridge engineering technology of the 1870s rather than the early 20th century. The span remains in service carrying local vehicular traffic over the Little Osage River. NR.

■ ■ ■ ■ ■ ■ MICHIGAN ■ ■ ■ ■ ■ ■ ■

ALLEGAN

The Second Street Bridge has served local transportation needs in Allegan for more than 100 years. Recently, it
took on even greater significance as an important symbol
of what local officials can do to save historic bridges if
they have the backing of politicians as well as state and
federal highway officials (see page 66). In a pioneering
rehabilitation project, the city of Allegan completely
disassembled the pin-connected, double-intersection
Pratt through truss and replaced its deteriorated vertical
compression members with similarly designed new steel
components. After reassembly, the refurbished truss was
then moved back on its original site by a sophisticated
trestle and roller system. Today, the 150-foot-long structure once again carries local traffic. NR, ASCE.

■ Second Street Bridge
Across the Kalamazoo River
On Second Street
King Iron Bridge and
Manufacturing Company
1886

Second Street Bridge after its
refurbishment in 1983.

CROTON

Although the flat terrain of the upper Midwest would not
appear well suited to the development of hydroelectric
plants, the region's relatively heavy amount of precipitation throughout the year has prompted construction of
numerous low-head power plants. Although not large in
comparison with plants in mountainous areas or along
major rivers, these smaller plants in the Midwest
historically played a prominent role in the growth of
regional electric power systems. Based around low-head
dams, these facilities produced substantial quantities of
reliable power because of the constant water flow in the
region's rivers. Among the more noteworthy of these early
plants is the Croton Dam and its generating station in
west-central Michigan. Completed in 1908 by the Grand
Rapids–Muskegon Power Company, the 40-foot-high,
670-foot-long earthfill dam impounds water for a con-
tiguous 15,000-horsepower generating plant. Power
from the Croton Dam was originally transmitted over a
35-mile-long power line to Grand Rapids at a pressure of
110,000 volts. As noted by Charles Hyde in an inventory

■ Croton Dam
Across the Muskegon River
3 miles north of State
Route 82
William G. Fargo
1908

of historic engineering sites in Michigan, this represented the highest commercial voltage in the United States at the time it first began to operate. Today, the Croton Dam installation still serves as part of the Consumers Power Company system that provides electricity for much of Michigan.

DETROIT

■ Ambassador Bridge
Across the Detroit River
Near State Route 3, between
Detroit and Windsor,
Ontario, Canada
Jonathan Jones
1929

Night view of the
Ambassador Bridge, with
lights on the cables.

Begun in 1927, the Ambassador Bridge remains in active use as a major example of suspension-bridge technology. Because of the huge amount of shipping on the Great Lakes, it was necessary that the bridge provide wide clearances to accommodate traffic on the Detroit River. Built by the McClintic-Marshall Company of Pennsylvania, the bridge has a clear span between towers of 1,850 feet and a total length, including approach spans, of more than 9,000 feet. For a brief period, before the opening of the George Washington Bridge (1931) over the Hudson River, it could claim credit as the longest-span bridge in the world. Today, the Ambassador Bridge is still a major commercial artery between the United States and Canada; it remains privately owned and operated.

REDRIDGE

■ Redridge Steel Dam
Across the Salmon
Trout River
10 miles west of State
Route 26
J. F. Jackson
1901

Built by the Atlantic and Baltic mining companies to supply water for their stamping mills, the Redridge Steel Dam replaced and inundated the original 1894 timber crib dam at the site. Located in Michigan's copper country on the north shore of the state's Upper Peninsula, the new structure was the second steel dam built in the United States. Fabricated by the Wisconsin Bridge and

Iron Company, the 1,006-foot-long, 74-foot-high steel structure rests on a large concrete foundation keyed into bedrock. As documented by engineering historian Terry Reynolds, the upstream face of the dam consists of steel plates that vary in thickness from three-eighths to three-sixteenths of an inch. The plates are inclined at a 45-degree angle into the reservoir. The upstream face is supported by steel I beams that rest directly on the concrete foundation. In essence, it is a flat-slab buttress design in which all the main components, except the foundation, are steel. Because mining operations are shut down, the dam is no longer used for water storage. Aside from a minor amount of surface scale, the dam is generally in excellent condition, especially given that it has experienced more than 85 years of harsh winters near the shore of Lake Superior. However, in the late 1970s a few portions of the steel upstream facing were removed in order to prevent water from impounding in the reservoir. Despite this, the dam retains most of its original design integrity.

This structure is a well-preserved example of timber crib dam technology dating to the late 19th century. Built by the Atlantic Mining Company to provide water for its nearby stamping mill on the shore of Lake Superior, the dam is 228 feet long at the crest with a maximum height of 53 feet. In 1901 it was replaced by a taller structure, the Redridge Steel Dam, directly downstream. For many years the timber crib dam lay submerged under the reservoir formed by the larger dam. In the 1970s the steel dam was removed from service and its reservoir permanently drained. Since the emptying of the reservoir the old timber dam has been visible to visitors who make their way to the upstream side of the steel dam. Although submerged for many years, the timber dam still possesses much of its structural integrity.

Redridge's steel dam (above), which took over the function of the timber crib dam (below) in 1901. Both dams are no longer in use.

■ **Redridge Timber Crib Dam**
Across the Salmon Trout River
10 miles west of State Route 26
1894

David Steinman's triumphal suspension bridge across the Straits of Mackinac.

ST. IGNACE

■ **Mackinac Bridge**
Across the Straits of Mackinac
On Interstate 75
David B. Steinman
1957

Located along the southern shore of Lake Superior and encompassing more than 13,000 square miles, the Upper Peninsula of Michigan is a sparsely populated region separated from the rest of the state by the Straits of Mackinac between Lake Michigan and Lake Huron. Plans to build a permanent crossing at the straits were developed in the 1930s, but nothing substantial occurred until the formation of the Mackinac Bridge Authority in 1950. Under its direction, David B. Steinman (1886–1960) developed plans for a huge steel suspension bridge with a central clear span of 3,800 feet. Including all approach spans, the entire structure stretches out for more than 17,900 feet. Construction began in the summer of 1954, and during the next four years more than 10,000 people worked on the project. First opened for traffic in November 1957, the Mackinac Bridge was officially completed in September 1958. It continues to serve as a vital transporation link between the Upper Peninsula and the southern part of the state. NR.

SMYRNA

■ **White's Covered Bridge**
Across the Flat River
Near State Route 91,
southwest of Smyrna
Jared N. Bresee and
J. W. Walker
c. 1867

This single-span, 119-foot-long wooden covered bridge is a rare example of a Brown truss. The Brown truss appears similar to a Howe truss except that it does not include any iron tension members running vertically between the top and bottom chords. Built in a state that once supported a huge lumber industry, White's Bridge is one of only three covered bridges that survive in Michigan. It is now preserved as a historic site and is accessible to pedestrians. NR.

VICTORIA

Located in the remote western reaches of Michigan's Upper Peninsula, the Victoria Dam is the tallest reinforced-concrete, multiple-arch dam in both the midwestern and northeastern United States. Rising 115 feet above its foundations, the main section of the dam consists of four semicircular arches supported on buttresses spaced 68 feet apart. Originally built by the Copper Range Company as part of a 15,000-kilowatt hydroelectric plant that served its mining facilities in the region, the Victoria Dam and power plant came under the control of the Upper Peninsula Power Company in 1947. The dam's isolated location represented an ideal setting for employment of a multiple-arch design because of the great cost of hauling construction materials to the site. This design minimized the amount of concrete required for the structure and thus dramatically reduced transportation costs associated with the project. Despite the long, harsh winters in the Upper Peninsula, the Victoria Dam continues to operate and plays a significant role in the region's electrical power grid.

■ **Victoria Dam**
Across the Ontonagan River
On Victoria Dam Road,
5 miles southwest of U.S.
Route 45
Holland, Ackerman and
Holland
1931

■ ■ ■ ■ ■ ■ ■ MINNESOTA ■ ■ ■ ■ ■ ■ ■

DULUTH

In the late 19th century Duluth became an important shipping center for iron ore. Located at the western edge of Lake Superior, the city's harbor was a critical nexus where trains filled with iron ore from the Mesabi Range met ships that would carry the ore to furnaces in the East. The original Aerial Bridge was built to provide access across the entrance of Duluth Harbor for pedestrians and small vehicles. Similar in design to several "transporter" bridges built in Europe, the structure supported a small gondola-like platform that moved back and forth across the channel. In essence, it functioned much like a ferry, except that the platform hovered a few feet above the water and was supported on cables attached to the top of

■ **Aerial Lift Bridge**
Across the Duluth Ship
Canal
On Lake Avenue
Thomas F. McGibray and
C. A. P. Turner
1905, 1929

Original Duluth transporter bridge, c. 1910. Note the movable platform hovering above the water.

the bridge. In 1929 the bridge was drastically modified, and the gondola mechanism was entirely removed. In its place C. A. P. Turner designed a standard type of vertical-lift bridge with a clear span of more than 350 feet. The new structure could carry a much greater amount of vehicular traffic across the channel, while also providing all the navigational clearance demanded by shipping interests. The modified bridge is still in operation. NR.

MENDOTA

■ **Mendota Bridge**
Across the Minnesota River
On State Route 55
C. A. P. Turner and
Walter H. Wheeler
1926

The Mendota Bridge is located only a few hundred yards upstream from the confluence of the Minnesota and Mississippi rivers. As such, it still plays an important role in the highway system serving the Minneapolis–St. Paul area. The bridge is 4,119 feet long and consists primarily of 13 open-spandrel, reinforced-concrete arches. Each of the main arches is 304 feet in length. As historian Nicholas Westbrook noted in a guide describing local industrial sites, the bridge's construction involved the use of steel centering arches to support the concrete formwork. After the concrete in an arch had been given sufficient time to harden, the steel supports were disassembled and reused for the construction of other arches. This technique reduced construction costs for the extremely long structure. The bridge remains in operation carrying highway traffic. NR.

MINNEAPOLIS

■ **Cappelen Memorial Bridge**
Across the Mississippi River
On Franklin Avenue
Frederick W. Cappelen and
Kristoffer Oustad
1923

Centered around the upper Mississippi River, the Twin Cities, Minneapolis and St. Paul, have become the site of many large, impressive bridges. Frederick W. Cappelen moved to the region in the early 1880s, and during the next 40 years he helped design and build several important railroad and highway spans. Shortly after World War I he began work on a 1,054-foot-long, reinforced-concrete arch bridge with a center span of 400 feet. At the time of construction this was the longest concrete arch in the world. Cappelen died before its completion, but his assistant Kristoffer Oustad oversaw the final stages of construction. In honor of Cappelen's work in developing the Twin Cities transportation system, the structure was named after him as a memorial. In 1971 the bridge underwent considerable renovation, with the deck and many of the spandrel supports being completely rebuilt. The arches still retain their historical integrity, however, and the bridge continues to carry large amounts of modern highway traffic. NR.

■ **James J. Hill Stone
Arch Bridge**
Across the Mississippi River
At St. Anthony Falls,
near U.S. Route 52
Charles C. Smith
1883

In the late 1870s James J. Hill and a group of business associates took over a small Minnesota railroad and, after renaming it the St. Paul, Minneapolis and Manitoba Railroad, began an ambitious plan to build a major transportation empire. In 1883 a subsidiary of this railroad constructed this huge 2,490-foot-long stone arch bridge across the Mississippi River in Minneapolis, which provided a permanent and visually impressive crossing. Later named in Hill's honor, the double-track structure incorporated 23 arches into its design and extended 77 feet above normal water level. In the 1890s it became a centerpiece of Hill's Great Northern Railroad. The bridge remained in use for almost 100 years, until the Burlington Northern retired it in 1982. Aside from the replacement of two arches with steel trusses to provide navigational clearance, the bridge is essentially un-altered. Although no longer used by the railroads, there are no plans to demolish the most prominent transporta-tion structure in the Twin Cities area. NR district. ASCE.

Above: James J. Hill Stone Arch Bridge, Minneapolis, carrying the Great Northern Railroad, the Empire Builder, c. 1910. Left: Dam No. 1 across the Mississippi River, with navigation locks in the foreground.

ST. PAUL

Soon after the Civil War ended in 1865, the Army Corps of Engineers began work on improving the navigability of the Mississippi River. At first this project primarily involved the removal of snags in the stream bed and the construction of wing dams to concentrate river flow in the deepest parts of the channel. But by the early 20th century, the corps began building permanent dams across the complete length of the river to provide for slackwater pools. Among the largest of these early corps

■ **Twin Cities Lock and Dam No. 1 (Ford Dam)**
Across the Mississippi River
Adjacent to the Ford
Parkway Bridge
U.S. Army Corps of
Engineers
1917

Dam No. 1, with the locks on the left and its hydroelectric plant on the right. The Ford factory is located to the right of the powerhouse.

structures was Lock and Dam No. 1, between Minneapolis and St. Paul. It is a reinforced-concrete, overflow Ambursen flat-slab dam, approximately 40 feet tall and 574 feet long. Construction began in late 1910 but, because of delays caused by the failure of temporary cofferdams (a technology designed to protect the foundations during construction), it was not completed until 1917. At the west end of the dam are two navigation locks operated by the corps. At the east end is a hydroelectric plant that provides power to the nearby Ford automobile factory. Because of this latter association, the dam is often referred to as the Ford Dam.

WINONA

■ **Winona Railroad Bridge**
Across the
Mississippi River
Adjacent to Minnesota
Route 43
George S. Morison
1891

Located approximately 100 miles downstream from St. Paul, Winona was recognized as a desirable spot for a permanent crossing of the Mississippi River as early as 1866. Nothing substantive on this matter occurred until the 1880s, however, when the Chicago, Burlington and Quincy Railroad sought to expand its operations into the Minneapolis–St. Paul area. The railroad reached St. Paul in 1886 via a rail line along the east (Wisconsin) side of the Mississippi River. A connection across the river at Winona was soon made by a ferry capable of carrying locomotives and rolling stock. But the ferry quickly proved cumbersome, and in early 1888 the Burlington, in concert with the Winona and Southwestern and the Green Bay and Western railroads, formed the Winona Bridge Railway Company. Chartered by the Wisconsin and Minnesota state legislatures as well as by Congress, this company existed only to build and operate the bridge; it maintained fewer than 5,500 feet of track. The Winona Bridge Railway Company was authorized to collect tolls from the other railroad companies using the bridge, but federal legislation required that the amount of these tolls be regulated by the secretary of war (at this time, transportation matters often were under the jurisdiction of the war department) to ensure that they were not excessive. The Winona Bridge was the first interstate bridge subject to such federal regulation.

In 1890 the bridge company contracted with the Union

Bridge Company to fabricate and build the span, and this firm retained George S. Morison (1842–1903) to prepare the design. Morison designed a five-span structure consisting of four fixed-span Parker through trusses and a single movable Pratt through swing truss. The 440-foot-long movable span provided clearance for navigation, a requirement of the federal charter authorizing construction. Work on the 2,640-foot-long bridge, which included more than 1,000 feet of approach spans, began in the late summer of 1890, and the first train rolled across it on July 4, 1891. It has remained in use since then with relatively little alteration.

Swing span of George Morison's Winona Railroad Bridge. The structure is in a closed position, ready for traffic.

■ ■ ■ ■ ■ ■ ■ MISSOURI ■ ■ ■ ■ ■ ■ ■

KANSAS CITY

The A.S.B. Bridge is a unique example of a double-decker, vertical-lift bridge built to carry both railroad and highway traffic. As originally constructed, the bridge carried railroad traffic on its lower level while automobiles and electric interurban cars traveled across the top. To facilitate commercial navigation along the Missouri River, the 428-foot-long lift span can be raised to provide clearance for barges. The structure is designed so that the lift span "telescopes" into the truss that supports the highway deck. Consequently, raising the

■ **Armor, Swift, Burlington (A.S.B.) Bridge**
Across the Missouri River
On the Burlington
Northern Railroad
right-of-way
Waddell and Harrington
1911, 1982

Armor, Swift, Burlington Bridge showing how the lower truss telescopes into the upper truss, c. 1920.

bridge disrupts only railroad service; highway traffic can continue undisturbed. In 1982 the Missouri Highway and Transportation Department removed the roadway from the top of the bridge, and the structure is now used exclusively by the Burlington Northern Railroad. The lift span continues to operate using its unique telescoping design.

The history of the A.S.B. Bridge dates to the late 1880s, when J. A. L. Waddell (1854–1938) first received a design commission from the Wabash, St. Louis and Pacific Railroad Company. At that time several masonry piers were built but construction soon halted, presumably a result of the financial panic of 1893. Financing remained in limbo for several years until the site came under the control of the Burlington Railroad and the Armour and Swift meat packing companies. In 1907 Waddell was asked to prepare a new design for the bridge, and the result is the present structure completed in December 1911. Waddell practiced as a consulting bridge engineer based in Kansas City from 1887 through 1920, and it is appropriate that his professional hometown be the site of one of his most distinctive designs. At times the A.S.B. Bridge is called either the Winner or the Fratt Bridge in honor of two turn-of-the-century Kansas City railroad executives. But the most common appellation refers to the companies responsible for its actual construction.

LAKE OF THE OZARKS

■ **Bagnell Dam**
Across the Osage River
On County Road W,
3 miles west of U.S. Route 54
Stone and Webster
1931

Left: Downstream face of the Bagnell Dam with the overflow spillway in the foreground. Right: Sign greeting visitors to the Lake of the Ozarks, c. 1940.

The Ozark Mountains in Missouri are renowned as a vacation paradise, and for more than 50 years the Lake of the Ozarks has been a focus of the regional tourist trade. Although the lake affords many recreational opportunities, it was built for a much more prosaic purpose — hydroelectric power production. The lake is impounded by the Bagnell Dam, a 148-foot high, 2,543-foot-long, straight-crested concrete gravity structure built by the Union Electric Company of Missouri. Power generated by the privately financed and owned dam is transmitted at high voltages to St. Louis (about 120 miles east) and other parts of the Mississippi River Valley. The dam continues to provide for electric power generation and to impound one of the Midwest's most popular recreation lakes.

Following completion of the Bagnell Dam (1931) as part of a privately financed hydroelectric project, attention soon focused on developing the reservoir's recreation potential. In the 1930s the National Park Service oversaw construction of the Lake of the Ozarks Recreational Area (later called Lake of the Ozarks State Park), a public facility on the shores of the reservoir. As part of a relief work effort during the Depression, the Civilian Conservation Corps erected numerous buildings in the developing park and also built a highway to provide access from U.S. Route 54. This highway work included construction of a 30-foot-long masonry-faced, reinforced-concrete arch bridge across Patterson Hollow. Despite the use of reinforced concrete, the stone facing creates a rustic appearance that was considered more appropriate for a park setting. Named after a local resident, the McDaniels Bridge is representative of numerous small bridges built by the CCC during the 1930s. Although unremarkable in size or design, it is nevertheless significant as an example of a small-scale public works bridge project undertaken during the Depression. The McDaniels Bridge is still used to carry highway traffic in the park. NR.

■ **McDaniels Bridge**
Across Patterson Hollow
In Lake of the Ozarks State
Park, on State Route 134
Civilian Conservation Corps
c. 1938

Masonry facade of the
reinforced-concrete
McDaniels Bridge, located in
a wooded setting.

PARKVILLE

In 1894 J. A. L. Waddell patented one of the most visually distinctive types of bridge designs of the late 19th century. Derived from earlier wooden king-post trusses, the Waddell "A" truss is a triangularly shaped design that employs pin connections to join together the main structural members. Waddell developed the design in an effort to find a way of building a rigid, short-span, pin-connected truss that could economically carry heavy, fast-moving railroad traffic without excessive deflection or vibration. The "A" truss achieves its structural stability because of its substantial height in the center of the structure. The top of the truss is tall enough to allow cross bracing to connect the two sides of the bridge and thus give the design great rigidity. In the 1890s scores of Waddell "A" trusses were built for railroads in the Midwest and in Japan (Waddell taught engineering in Tokyo during the 1880s, and he had close professional ties to this island nation). With the widespread deploy-

■ **Waddell "A" Truss Bridge**
Across Rush Creek in
English Landing Park
½ mile south of State
Route 9
J. A. L. Waddell
1898; reerected 1987

Waddell "A" Truss Bridge, before it was moved to Parkville.

ment of field-riveting technology in the early 20th century, the market for pin-connected "A" trusses disappeared, but not before the design proved its utility for numerous short-span railroad crossings.

The "A" truss in Parkville was originally built in 1898 by the Quincy, Omaha and Kansas City Railroad on its line between Kansas City and Pattonsburg, Mo. Located over Linn Branch Creek near the town of Trimble, about 30 miles north of Kansas City, the bridge carried rail traffic until the line was abandoned in the 1930s. After World War II the bridge was adapted for use on a local highway and it remained in service until threatened with inundation by the U.S. Army Corps of Engineers' Smithville Dam. In 1982 the corps disassembled the Waddell "A" truss over Linn Branch Creek and stored the structural members, hoping to reerect it at some other location. This opportunity came in 1987 when the city of Parkville obtained the span from the corps (for the grand sum of $200 to cover paperwork costs) for use as a pedestrian bridge in English Landing Park near the banks of the Missouri River. The planning and implementation of the bridge's reconstruction involved a broad range of hard-working participants, including Prof. George Hauck's systems design class at the University of Missouri–Kansas City, Local 10 of the Iron Workers Union, the engineering firm of Howard, Needles, Tammen and Bergendoff, the Bratton Corporation steelworking group, the construction firm of Wilkerson-Maxwell Company and the Parkville City Park Board. The 100-foot-long, 40-foot-high bridge was formally dedicated at its new site in November 1987, and it is now ready to serve the needs of pedestrians enjoying the pleasures of Parkville's riverfront park.

ST. LOUIS

■ **Eads Bridge**
Across the Mississippi River
At Washington Street
James B. Eads
1874

Capt. James B. Eads (1820–87) started his career as a salvager, diving into sunken ships along the Mississippi River and retrieving whatever valuable cargo they might hold. When Eads began promoting construction of a large railroad bridge across the Mississippi River after the Civil War, he did so not as an experienced structural engineer, but as someone with an unparalleled knowledge of the river's bottom who could best figure out how to

build permanent piers on the floor of the treacherous, silt-laden waterway. With dreams of establishing St. Louis as a major center of railroad service in the Midwest, Eads developed a three-span, steel deck arch bridge design that was unprecedented in size in American engineering practice. Because of its dimensions, the design drew criticism from other bridge engineers. However, the force of Eads's personality and his belief in the project ultimately convinced the financial backers of the Illinois and St. Louis Bridge Company to appoint him chief engineer and approve the triple-arch structure.

Construction on the all-important piers began in the summer of 1869, and, despite the effect of caissons disease (the bends) on workers, the masonry piers were completed in 1873. Erection of the arches using a cantilever method of construction began soon afterward. The three-arch superstructure (two are 502 feet long and one is 520 feet long) was completed in 1874, and the bridge opened to railroad and highway traffic shortly afterward. Ironically, the bridge proved to be a great structural success, but it quickly became a financial albatross. Little railroad traffic passed over the structure, and, without the anticipated toll revenue, the bridge owners went bankrupt within a year. Today, the structure carries only highway traffic, but the aesthetic power of Eads's arch design has not been diminished. The bridge is considered by historians to rank with the Brooklyn Bridge as one of the most significant in the United States. It is distinguished by the early use of steel in the design, the unprecedented employment of pneumatic (pres-

Eads Bridge. Top left: Masonry arch approach spans near the Illinois shore. Top right: Inside an arch, showing the arch's crown and abandoned railroad trackage. Above: Bridge's three main arch spans, with the Gateway Arch in the background.

surized) caissons to help build the foundations, and the graceful play of the huge arches that contrasts with the monolithic masonry piers. Although Eads never built another bridge (he spent much of his later career advocating the construction of jetties to keep the mouth of the Mississippi River clear for deep water navigation), the monumental structure he erected at St. Louis is a true landmark in the annals of American engineering. NR, ASCE.

WEST ALTON

■ Bellefontaine Bridge
Across the Missouri River
Adjacent to U.S. Route 67
George S. Morison
1893

In the early 1890s the Chicago, Burlington and Quincy Railroad sought access to St. Louis from its main line to the north. Acting through a subsidiary, the St. Louis, Keokuk and North Western Railroad, the Burlington line crossed the Mississippi River at Alton, Ill., a city about 15 miles north of St. Louis. This crossing was also about four miles upstream from the confluence of the Mississippi and Missouri rivers. Consequently, the railroad also had to build a major bridge across the latter. Designed by George S. Morison (1842–1903), a prominent engineer responsible for several other Burlington structures, the Bellefontaine Bridge is a four-span, all-steel, pin-connected Baltimore through truss. Each span is 440 feet long, and the overall structure, minus approach spans, is more than 1,700 feet long. Excavation for the pier foundations began in early July 1892, and construction of the piers continued for a year. Erection of the steel superstructure occurred at a much faster pace, reflecting the speed possible in assembling pin-connected trusses. Work on the trusses began in mid-September 1893 and was completed by mid-December. More than 90 years old, the Bellefontaine Bridge still carries railroad traffic serving St. Louis.

Bellefontaine Bridge. Left: Portal entrance and the nameplate crediting the design and construction engineers. Right: Subdivided Baltimore truss supported on masonry piers.

■ ■ ■ ■ ■ ■ NEBRASKA ■ ■ ■ ■ ■ ■ ■

HEMINGFORD

In the early 20th century Nebraska farmers in the North Platte River Valley benefited from the construction of the U.S. Reclamation Service's Pathfinder Dam (1910) in Alcova, Wyo. In addition, the service built a few small distribution reservoirs in the state. It was many years, however, before the Bureau of Reclamation undertook construction of a large storage dam within the borders of Nebraska. Finally, in 1941 workers hired by the Works Progress Administration began building the bureau-designed Box Butte Dam as part of the Mirage Flats Project in northwest Nebraska. The dam, an 87-foot-high earthfill structure with a total length of 5,508 feet, provides water to land under the jurisdiction of the Mirage Flats Irrigation District. Most of the district was originally part of the large Peters-Williams ranch, which, beginning in the 1920s, was broken up into smaller tracts and sold to individual farmers. The dam continues to play a significant role in the economy of northwestern Nebraska.

■ **Box Butte Dam**
Across the Niobrara River
10 miles north of State
Route 2
U.S. Bureau of Reclamation
1944

OGALLALA

When early pioneers trekked along the Oregon Trail from Independence, Mo., to the Pacific Northwest, much of their initial journey followed the Platte and North Platte rivers. At first, settlers saw this part of the Great Plains as a region to travel through, not a destination unto itself. But by the late 19th century, homesteaders were attracted to the area as a place to live, work and, with a bit of luck, prosper. Of course, the major impediment to economic development was the lack of water. To supplement the natural accumulation of rain and snow, some farmers in the Platte and North Platte river valleys were able to divert water for irrigation. This irrigation development remained relatively small-scale, however, until the construction of the Kingsley Dam.

Completed with financing from the Central Nebraska Public Power and Irrigation District, the dam is an earthfill structure 162 feet tall and more than 15,500 feet long. When built it was the second largest dam of its type in the United States, after the Fort Peck Dam (1940) in Montana. The Kingsley Dam has a storage capacity of two million acre-feet of water, used to irrigate 138,000 acres of arable land south of the Platte River (under the jurisdiction of the Central Nebraska Platte Power District) and 60,000 acres of arable land north of the Platte River (under the jurisdiction of the Nebraska Platte Power District). The dam also benefits several smaller irrigation districts to the east of Ogallala. Corn and to a lesser extent soybeans, milo and alfalfa are the primary crops nourished by the dam's water. Still in full operation, the dam is a vital part of south-central Nebraska's economy. Its reservoir, named Lake C. W. McConaughy, is

■ **Kingsley Dam**
Across the North Platte
River
On State Route 61
George E. Johnson
1931

also a major recreational attraction in the region. The dam and reservoir are named after George P. Kingsley and Charles W. McConaughy, two prominent advocates of the project during the early 20th century.

■ ■ ■ ■ ■ ■ NORTH DAKOTA ■ ■ ■ ■ ■ ■

MINOT

■ Eastwood Park Bridge
Across the Souris River
On Central Avenue
T. W. Sprague
1927

In 1927 the Ward County board of commissioners acted to authorize construction of a bridge connecting Minot's downtown business district with the residential area of Eastwood Park. The county specified an arch bridge but, apparently because of fears that such a design might infringe on patents held by the Marsh Engineering Company, an alternative design was prepared by T. W. Sprague of the North Dakota highway department. Sprague's design appears to be a reinforced-concrete rainbow arch bridge but is, in fact, a cantilever structure. The arch that rises above the center part of the span has no structural function and was included only for visual effect. The 144-foot-long bridge was removed from highway use in the mid-1970s and adapted for pedestrian use. It survives today as an example of bridge design in which nontechnical and aesthetic concerns played a key role in the final design. NR.

Eastwood Park Bridge, which looks like an arch but actually functions as a reinforced-concrete cantilever structure.

RIVERDALE

■ Garrison Dam
Across the Missouri River
Adjacent to State Route 200
U.S. Army Corps of
Engineers
1954

The Army Corps of Engineers did not lose interest in building dams along the Missouri River following completion of Montana's Fort Peck Dam in 1940. In fact, after the devastating flood of 1943, the U.S. House of Representatives directed the corps to undertake a complete review of flood-control programs along the river. Known as the Pick Report, for Col. Lewis Pick, the corps' subsequent study outlined a far-ranging program incorporating numerous dams for flood control, hydroelectric power, navigation and irrigation. The Pick Report covered the entire Missouri River watershed and, in so doing, encroached on the U.S. Bureau of Reclamation's plans for projects in the upper river valley. Eventually the two federal agencies developed a compromise program called the Pick-Sloan Plan, named in part for William Sloan, an assistant director of the bureau, which became the blueprint for all later federal water-resource development along the river.

The Garrison Dam was the first major dam built by the corps in line with the Pick-Sloan Plan. Construction of the 200-foot-high, 11,200-foot-long earthen dam began in 1946 and ended eight years later. By 1956 the first electric power generators at the dam were in operation. Although the Garrison Dam was originally designed to provide hydroelectric power, flood control and irrigation, in general only the first two of these functions have been fulfilled. Plans to pump water from the dam's reservoir, known as Lake Sakakawea, to eastern North Dakota remain unimplemented, although federal funding for this irrigation project continues to be a focus of congressional debate. The reservoir has also become notorious because it flooded out the ancestral home of the Arikara Indians, a prominent Native American tribe.

■ ■ ■ ■ ■ ■ ■ ■ OHIO ■ ■ ■ ■ ■ ■ ■ ■

BALTIMORE

While wood and iron combination bridges were common in the 19th century, the John Bright Covered Bridge is an unusual example of this technology. Named after an early pioneer in the region, it is 74 feet long and consists of hybrid trusses that superimpose a wooden arch against an iron suspension chain. The structure is the only example of its type known to exist in the United States. It still carries local traffic loads. NR.

■ **John Bright**
Covered Bridge
Across Poplar Creek
On Township Road 263
August Borneman and Sons,
builder
1881

BLADEN

**■ Gallipolis Lock
and Dam**
Across the Ohio River
Adjacent to State Route 7
U.S. Army Corps of
Engineers
1937

Commercial navigation along the Ohio River began in the early 19th century and became the source of numerous folk tales about Mike Fink and other keelboatmen who plied the waters between Pittsburgh and New Orleans. With steamboats also came a strong desire to remove shoals, rapids and other impediments to travel along the river. In the late 1870s the Army Corps of Engineers began work on the Davis Dam with the intention of building a slackwater pool to facilitate boat movement within Pittsburgh's harbor. This structure became the first of several movable dams along the Ohio River built by the corps to provide a deep, reliable channel for inland navigation.

In the 1930s the corps received authorization to construct an even deeper channel, prompting construction of several new dams. Among the most impressive of these was the Gallipolis Lock and Dam, a 41-foot-tall, 1,225-foot-long roller dam. The eight steel roller sections of the structure can be raised in times of high water to allow floods to pass by quickly. During periods of normal or low water, the rollers are lowered and allow water to be impounded behind the structure. Because of the Gallipolis Dam and other similar corps structures, the Ohio River is still a major inland navigation route.

Piers and roller sections of the Gallipolis Dam. During high water the rollers move up and allow more water to flow between the piers.

CINCINNATI

**■ Cincinnati Suspension
Bridge**
Across the Ohio River
On Kentucky State Route 17
John A. Roebling and
Wilhelm Hildenbrand
1867, 1899

The Brooklyn Bridge is usually considered John A. Roebling's greatest accomplishment. But before undertaking the design of a suspension bridge across the East River in New York City, Roebling (1806–69) worked on several large projects that contributed to his knowledge of long-span bridge technology. Among the most important of these was this 1,057-foot-long clear span bridge across the Ohio River between Cincinnati and Covington, Ky. Work on the masonry towers began in 1856, but financial problems associated with the panic of 1857 soon caused construction to halt. Work resumed in 1863, and the bridge finally opened for traffic in early 1867. The structure provided exemplary service throughout the 19th century, but by the 1890s local authorities were concerned that the deck structure was not adequate to

carry increasing traffic loads. In 1899 a new steel truss designed by Wilhelm Hildenbrand, a former Roebling draftsman who worked on the Brooklyn Bridge, replaced the original deck and substantially altered the appearance of Roebling's design. The new deck allowed the bridge to accommodate 20th-century highway traffic, however, and the structure still remains in active use. NR, ASCE.

Shortly after construction in 1889 of America's first reinforced-concrete bridge, the Alvord Lake Bridge (1889) in San Francisco, the technology began to be promoted throughout the United States. Although not necessarily cheaper than comparable iron or steel spans, reinforced-concrete bridges offered opportunities to design structures more in keeping with contemporary Beaux Arts architectural fashion. Significantly, many early reinforced-concrete bridges were built in parks where their arch designs often recalled classical architectural motifs.

Among the earliest of these structures is the 137-foot-long Eden Park Bridge. Designed and built by Fritz von

Cincinnati Suspension Bridge. Top: Roebling's original bridge, c. 1875. Above left: Bridge from the Kentucky shore. Above: Detail of the steel deck truss added in the 1890s.

■ **Eden Park Bridge**
Across Eden Park Drive
In Eden Park, adjacent to
the watertower
Fritz von Emperger
1895

Top: Eden Park Bridge shortly after completion, c. 1900. Right: Recent view showing its new balustrade.

Emperger using a technique developed by his fellow Austrian Josef Melan, the Eden Park Bridge is an example of a Melan arch bridge. In this type of structure, the steel reinforcement is provided by a series of curved I beams placed near the bottom of the arch. Concrete was then poured around the I beams to hold them in place within the structure. The Melan design was used extensively only for a few years, largely because it did not make efficient use of its reinforcement in resisting shear stresses, but it played an important role in establishing the viability of reinforced-concrete bridge construction. The Eden Park Bridge recently underwent a comprehensive rehabilitation and is now ready for its second century of service.

CLEVELAND

■ Center Street Bridge
Across the Cuyahoga River
On Center Street
King Bridge Company
1901

The heart of Cleveland's historic industrial district is the Cuyahoga River, a tributary of Lake Erie. The river developed as a major transportation corridor for ships serving a variety of industries, including iron and steel furnaces. To facilitate vehicular traffic across the river while also accommodating navigational interests, numerous movable bridges were built by the city and railroad companies. In the 19th century at least 20 swing bridges, none of which survives, were built across the Cuyahoga River in Cleveland for highway and railroad use.

As documented by Cleveland historian Carol Poh Miller, the 249-foot-long Center Street Bridge is the oldest swing span in the city. Fabricated by Cleveland's own King Bridge Company, which was established by Zenas King, an early innovator in movable-bridge technology, the Center Street Bridge is a relatively rare example of a bob-tail swing bridge. In a bob-tail span the circular support for the structure is not located between

Center Street Bridge. Top:
Bridge in its closed position.
Above: Structure swung open
to allow passage of a
freighter. Left: Detail of the
rim-bearing support.

two arms of equal length. Instead, the part of the bridge
that extends out over the river is much longer than the
shore arm. To keep the bridge balanced the shore arm is
supplemented by a steel or concrete counterweight built
into the deck. Normal swing bridges usually require
construction of a pier in the middle of the river, thus
limiting the size of the passageway available for ships.
The use of a bob-tail design allowed the circular support
to be built on the river bank and thus not obstruct
navigation. Originally powered by a steam engine, the
Center Street Bridge is now run by electric motors. It
remains in active use.

■ Detroit-Superior High Level Bridge

Across the Cuyahoga River
At Detroit and
Superior Avenues
Frank R. Lander, A. M.
Felgate, W. A. Stincomb and
A. W. Zesinger
1918

Movable bridges reduced but did not eliminate transportation problems in the Cuyahoga River Valley; when a movable bridge opened to allow passage of a ship, vehicular traffic necessarily came to a stop. To help relieve this problem, the city built its first fixed-span, high-level bridge across the valley along Detroit and Superior avenues in 1918. This structure provides a 93-foot-high clearance over the river and thus eliminated the need for a movable span. The overall bridge is 3,112 feet long and includes 12 reinforced-concrete arches and a 591-foot-long, three-hinged steel arch over the river. It was built with two decks, the top for automobiles and the bottom for streetcars. Today, only the automobile deck is still used to carry traffic. NR.

Detroit-Superior High Level Bridge. Right: Bridge, c. 1920. Below left: Detail of an abutment hinge in an arch. Below right: Multirib deck arch span. Bottom: Steel through arch, with reinforced-concrete arch approach spans.

DRESDEN

From 1853 through 1913 a large wrought-iron suspension bridge with masonry towers carried highway traffic across the Muskingham River at Dresden. In 1913 devastating floods swept away the old span and forced construction of a new bridge. This replacement 705-foot-long steel suspension bridge has a clear span between towers of 443 feet. Instead of using cables for the tension members, the structure utilizes eyebar chains held together by steel pins. The traffic deck is a riveted Warren pony truss. The bridge still carries regional highway traffic. NR.

■ **Dresden Suspension Bridge**
Across the Muskingham River
On State Route 208
Bellefontaine Bridge and Steel Company
1914

Dresden Suspension Bridge, with eyebar suspension cables and Warren pony truss traffic deck.

ENGLEWOOD

In March 1913 heavy rainstorms wreaked havoc in southwestern Ohio by flooding out many towns and destroying numerous bridges, buildings and commercial developments. This disaster prompted the formation of the Miami Conservancy District to provide flood control for the area surrounding the city of Dayton. This district is noteworthy because it was among the first regionally based public works authorities formed to offer service and protection to citizens over a widespread geographical area. Born out of a Progressive-era desire to serve public needs through governmental action, the Miami Conservancy District is also remembered because its first chief engineer, Arthur E. Morgan (1878–1975), later went on to become the first board chairman of the Tennessee Valley Authority in the early 1930s. The Englewood Dam, an earthfill structure, is located about 10 miles northwest of Dayton and is designed to capture floods along the Stillwater River. Measuring 111 feet high and more than 4,700 feet long, it was the tallest flood-control dam built by the district in the early 1920s. The dam remains in service and continues to provide Dayton and the lower Miami River Valley with flood protection.

■ **Englewood Dam**
Across the Stillwater River
North of U.S. Route 40
Miami Conservancy District
1921

HOWARD

■ Kokosing River Bridge
Across the Kokosing River
On abandoned County Route
35, south of U.S. Route 36
Columbia Bridge Company
1874

The Kokosing River Bridge is a well-preserved example of a double-intersection Pratt through truss built by an important 19th-century Ohio bridge company. Established by David H. Morrison (1817–82) of Dayton, the Columbia Bridge Company flourished during the 1870s and early 1880s. Its Pratt truss designs were characterized by a patented type of compression member in which the wrought-iron channels were joined together with special iron "packing blocks" instead of standard riveted lattice work. The Kokosing River Bridge originally consisted of two spans, each 182 feet long, fabricated by Morrison's company. In 1913, however, one of the spans washed away and was replaced by a new Pratt truss. The entire structure has been bypassed by a new road and no longer carries highway traffic.

Right: Kokosing River Bridge, a Pratt through truss. Below: Drawings of David H. Morrison's patented double-intersection Pratt truss.

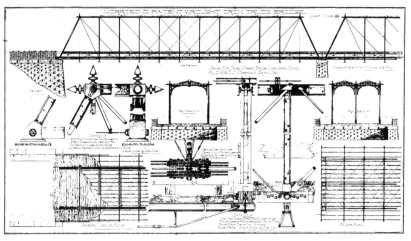

NEW CONCORD

■ S Bridge
Across Fox Creek
Adjacent to U.S. Route 40
1828

Construction of the National Road from Maryland to Ohio in the early 19th century involved building scores of masonry bridges across small creeks and rivers. Among the more visually distinctive of these structures were S bridges, constructed in locations where the highway crossed a stream at an oblique angle. To eliminate the need for building a lengthy skewed bridge at such sites, the roadway was contorted into an S shape that allowed

Sinuous curve of the S Bridge as it crosses Fox Creek near modern-day U.S. Route 40.

use of a relatively short-span arch across the river itself.

The S Bridge at New Concord is a good example of this structural form. It has a total length of 146 feet, but the single arch built perpendicularly across the stream is only 30 feet in span. The structure carried highway traffic for more than 100 years, but, with the advent of high-speed automobile travel, the S shape became a hazard. In 1936 the bridge at New Concord was bypassed during re-alignment of Route 40. Today, it is accessible to motorists as part of a roadside park. NR.

OREGONIA

The Oregonia Road Bridge, with a span of 212 feet, is one of the oldest and longest lenticular trusses in the United States. Still in highway use, it serves as striking evidence of how far afield the Berlin Iron Bridge Company of Berlin, Conn., was able to market its distinctive design. It is the only surviving lenticular through truss bridge in Ohio and continues to carry local vehicular traffic. It is also distinguished by ornate nameplates located over the portals at both ends of the truss.

■ **Oregonia Road Bridge**
Across the Little Miami River
On County Route 12, north of State Route 350
Berlin Iron Bridge Company
1883

POLAND

Founded as Range No. 1, Township No. 1, in Ohio's famous Western Reserve, the town of Poland became an important suburb of Youngstown in the late 19th and early 20th centuries. During this period iron and steel manufacturing developed into a key industry in the region. Not surprisingly, when the county replaced an existing wooden covered bridge on Cemetery Drive in 1877, it selected an iron bridge fabricated by a well-known Ohio bridge company. The design chosen was a wrought-iron bowstring arch truss patented by William Rezner, a physician who also dabbled in structural design. As documented by David Simmons, a historian specializing in Ohio bridge history, the Wrought Iron Bridge Company of Canton received the contract to build the 130-foot-long span over Yellow Creek based on Rezner's patent. Although the company also had its own patented bowstring design, this did not preclude its fabrication of designs developed by others. Rezner's design is distinguished by the use of ovoid tubular arches

■ **Yellow Creek Bowstring Arch Truss Bridge**
Across Yellow Creek
On Cemetery Drive, near the Riverside Cemetery
Wrought Iron Bridge Company
1877

for the top-chord compression members. No longer used to carry vehicular traffic, the Yellow Creek Bridge has been the cause of local concern as residents seek to find a way to preserve the structure.

STRUTHERS

■ **Lake Hamilton Dam**
Across Yellow Creek
Near State Route 616
David M. Wise
1907

The Youngstown area is best known for the extensive development of iron and steel mills that occurred there in the late 19th and early 20th centuries. In 1906, to increase the amount of water available for industrial and domestic purposes, the Youngstown Sheet and Tube Company financed construction of the Lake Hamilton Dam in the nearby town of Struthers. Built across a tributary of the Mahoning River, the structure is a 70-foot-high, 262-foot-long masonry gravity dam. A large spillway at the west end of the site prevents the dam from overtopping. Dams are often associated with industrial enterprises because of their role in hydraulic power development; however, the Lake Hamilton Dam is an unusual example of a water-impoundment structure built by industrial interests to serve more general water supply needs. Although the steel mills of Youngstown are now shut down, the dam remains an important part of the community's public works system. NR.

■ ■ ■ ■ ■ ■   SOUTH DAKOTA   ■ ■ ■ ■ ■ ■

BELLE FOURCHE

■ **Belle Fourche Dam**
Across the Belle Fourche
River
3 miles north of U.S.
Route 212
U.S. Reclamation Service
1911

Located to the north of the Black Hills near the western edge of South Dakota, the Belle Fourche Dam was authorized as one of the Reclamation Service's first dam projects. In the early 1890s Belle Fourche (which means "beautiful fork" in French) developed into a major livestock shipping point for the Chicago and North Western Railroad. The region appeared to offer strong

Opposite bottom: Overflow
spillway of the Belle Fourche
Dam. Left: Dam nearing
completion, 1911. Below: Bill
Olson Bridge, one of many
Pratt trusses built to serve
land made arable by the
Belle Fourche Dam.

irrigation possibilities, prompting the Reclamation Service to start building a large 122-foot-high, 6,262-foot-long earthfill dam at Belle Fourche in 1906. When completed five years later, it was one of the largest earthfill dams in the world. Water from the dam was delivered to the irrigation project lands through canals more than 40 miles long, and this lengthy delivery system caused engineering problems that affected the initial economic success of the Belle Fourche project. Today, however, the dam survives as a major component in western South Dakota's economic system.

Following completion of the Belle Fourche Dam in 1911, the region surrounding the new reservoir began to develop. Many bridges were built in the rapidly growing agricultural area, including the Bill Olson Bridge (named after a local farmer) across the Belle Fourche River. This is a 188-foot-long, single-span, pin-connected Pratt through truss with a wooden traffic deck. It is now maintained for local highway traffic by the Butte County highway department. NR.

■ **Bill Olson Bridge**
Across the Belle Fourche River
1 mile north of U.S. Route 212
Canton Bridge Company
1913

PICKSTOWN

Located only a few miles north of the Nebraska border, the Fort Randall Dam is the oldest major dam on the Missouri River in South Dakota. Designed for hydroelectric power production, the 160-foot-high, 10,700-foot-long earthen structure impounds a 150-mile-long reservoir. Construction of the dam required the creation

■ **Fort Randall Dam**
Across the Missouri River
Adjacent to U.S. Route 8
U.S. Army Corps of Engineers
1954

Fort Randall Dam, with its powerhouse in the center and spillway on the right.

of Pickstown, a new town capable of housing more than 3,000 workers, and also necessitated building a new railroad spur line to deliver material to the site. The Fort Randall Dam currently provides more than 300,000 kilowatts of power to the upper Missouri River Valley.

PIERRE

■ Oahe Dam
Across the Missouri River
Adjacent to State
Route 1806
U.S. Army Corps of
Engineers
1958

Army Corps of Engineers' earthfill Oahe Dam. Its powerhouse is on the right.

The Oahe Dam was the fourth major dam on the Missouri River built by the Army Corps of Engineers. In 1950, four years before completion of both the Garrison Dam (1954) at Riverdale, N.D., and the Fort Randall Dam (1954) at Pickstown, S.D., the corps began construction of the Oahe Dam. Located only six miles upstream from the state capital in Pierre, this is a 242-

foot-high. 9,300-foot-long earthen dam built between the high bluffs that rim the river bed. Shortly after work began, the dam became a focus of controversy as farmers expressed concern about the valuable farmland it would flood. Private power companies also cast aspersions on its need and practicality. In 1953 the newly installed Eisenhower administration pushed to cancel funding for the dam, but, because of strong congressional support, by 1954 the Oahe was again given complete federal approval. Final closure of the earthen embankment occurred in 1958, and hydroelectric power generation began in 1962. It remains in active service.

YANKTON

The Meridian Bridge was inspired and primarily financed by Yankton merchant Charles Gurney to provide a highway and railroad crossing over the Missouri River between South Dakota and Nebraska. It is a seven-span riveted Pratt deck truss extending a total of 1,668 feet. To facilitate river traffic, the second span from the west (South Dakota side) is a vertical-lift span. As documented by Junius R. Fishburne, the design originally called for an upper deck for highway traffic and a lower deck for rail traffic. The bridge, however, never served any railroads, and in 1953 the lower deck was converted to carry southbound traffic. In 1946 the city of Yankton purchased the bridge from Gurney and his associates. After paying off the bonds in 1953, the city turned the bridge over to the highway departments of South Dakota and Nebraska, which still have jurisidiction over its maintenance and operation. Although now supplemented by a highway across the Gavins Point Dam a few miles west of Yankton, the bridge continues to carry heavy daily interstate traffic. Because the Missouri River is no longer considered a navigable river at this location, the bridge's lift span is not maintained in operating condition.

■ **Meridian Bridge**
Across the Missouri River
On State Route 81
Charles Gurney
1924

Meridian Bridge's vertical-lift span, distinguished by the tall towers that once supported heavy counterweights.

■ ■ ■ ■ ■ ■ ■ WISCONSIN ■ ■ ■ ■ ■ ■ ■

CHIPPEWA FALLS

■ **Spring Street Bridge**
Across Duncan Creek
On Spring Street, near State
Route 124
James B. Marsh
1916

The Spring Street Bridge is a well-preserved example of a reinforced-concrete Marsh rainbow arch bridge, one of scores of these designs built in the Mississippi River Valley between 1910 and 1930. In 1916 James B. Marsh, the patent holder, personally visited the site of the Spring Street Bridge and proposed a rainbow arch design. This type of structure resembles a bowstring arch truss but employs reinforced concrete instead of steel or iron for its structural members. The Chippewa Falls city council approved the proposal, and in August of that year construction began. In October the bridge, which has a clear span of 90 feet, opened for vehicular traffic, and it has remained in operation ever since. NR.

Right: Spring Street Bridge showing how the reinforced-concrete floor beams are connected to the arch. Below: Roadway on the bridge, a Marsh rainbow arch design. Opposite top: Cedar River span, Wisconsin's only surviving covered bridge.

FIVE CORNERS

Although scores of covered bridges were built in this once heavily forested state, the 180-foot-long Town lattice truss over Cedar River is the only surviving example in Wisconsin. In 1962 it was removed from highway use and is now preserved as part of a county park. A concrete pier was added to support the structure at midspan, a common method of strengthening covered bridges, but otherwise it retains a substantial amount of its original structural integrity. NR.

■ **Cedar River Covered Bridge**
Across the Cedar River
On Covered Bridge Road,
near State Routes 143 and 60
1876

GALESVILLE

In the 1860s a ferry began operating across the Black River between Trempealeau and La Crosse counties in western Wisconsin. The ferry service proved adequate for several years but, with the growth of logging in the region, became difficult to maintain. As documented by engineering historian Bob Frame, in 1891 the La Crosse County board approved construction of seven bridges to carry highway traffic across the Black River lowlands. Built by the Clinton Bridge Company of Clinton, Iowa, all seven of these structures were metal bowstring arch trusses. In 1920 one of the original trusses became damaged and was replaced by a wooden king-post truss. In the 1950s the bridge system was taken out of service and acquired by the Wisconsin Department of Natural Resources as part of the Van Loon Wildlife Area. At this time the westernmost span of the group was demolished, but the remaining six bridges survive and are used to provide pedestrian access to the wildlife area. The Van Loon bridges, all within a mile of one another, include three two-span structures and three single-span structures varying in length from 50 to 141 feet. NR.

■ **Van Loon Wildlife Area Truss Bridges**
Across the Black River
In the Van Loon Wildlife Area
Clinton Bridge Company
1892, 1920

SOUTHWEST

Cave Creek Dam under construction in December 1922. Several public and private agencies financed this reinforced-concrete multiple-arch dam to protect Phoenix from floods.

■ ■ ■ ■ ■ ■ ■   ARIZONA   ■ ■ ■ ■ ■ ■ ■

CAMERON

**■ Cameron Suspension
Bridge**
Across the Little Colorado
River
Adjacent to U.S. Route 189
U.S. Office of Indian Affairs
and Midland Steel Company
1911

The Little Colorado River flows through a deep gorge in
northern Arizona before entering the Colorado River
from the south at a location upstream from the Grand
Canyon. The gorge made construction of midstream
piers impractical and prompted the design of a long-span
suspension bridge. The resulting Cameron Suspension
Bridge is an early example of large-scale suspension
bridge construction in the arid Southwest. It has a main
span of 660 feet and is distinguished by a Pratt through
truss supported by the suspension cables. The truss, in
turn, supports and stiffens the traffic deck. The bridge
remained in highway use for almost 50 years until it was
replaced in 1958. During this time it also carried
numerous herds of sheep being transferred to various
parts of the nearby Navajo Indian reservation. In 1958
the bridge was adapted to carry an oil pipeline and today
is part of an Atlantic Richfield Company system transporting crude oil from California to refineries in Texas.
NR.

CAVE CREEK

■ Bartlett Dam
Across the Verde River
On Bartlett Dam Road, 20
miles west of Cave Creek
U.S. Bureau of Reclamation
1939

The Verde River is a major tributary of the Salt River and
drains a large section of central Arizona. After a lengthy
battle over water rights, in the mid-1930s the Salt River
Valley Water Users Association joined with the Bureau of
Reclamation in constructing a dam on the Verde to help
protect the Phoenix area from floods and facilitate
increased irrigation. Known as the Bartlett Dam, this
structure is noteworthy because it is the only large
reinforced-concrete, multiple-arch storage dam ever built
by the bureau. At a height of 273 feet and with a crest
length of 900 feet, it is a good example of a double-walled,
multiple-arch dam in which the buttresses are thick,
hollow structures built to make the dam appear more
massive. It remains in active use as part of the water
supply system serving Phoenix and other communities in
the region.

Downstream side of the
Bartlett Dam, with its
spillway on the left. No strut-
tie beams were used to
connect adjacent buttresses.

In the West the major problem with water is that there is too little of it. But in certain areas at certain times, the problem is too much water — devastating floods that can wreak enormous damage to cities, railroads, highways and irrigation works. Before the 1920s the Phoenix region was susceptible to flash flooding along Cave Creek, a normally dry stream bed capable of carrying large quantities of floodwater from mountain areas north of the city. To eliminate this threat, the city of Phoenix joined with the Salt River Valley Water Users Association (now the Salt River Project), Maricopa County, the Santa Fe Railway and other parties to fund construction of a flood-control dam across Cave Creek. Original plans called for an earthfill structure. However, the Santa Fe Railway's experience in helping build an inexpensive, reinforced-concrete, multiple-arch dam (1918) at Lake Hodges in San Diego County, Calif., prompted consideration of a multiple-arch design for Cave Creek. A design of this type, prepared by John S. Eastwood (1857–1924), proved so economically attractive that it was chosen instead of an earthfill design.

Work on the 120-foot-high, 1,700-foot-long structure began in early 1922, and in March 1923 it proved its worth by holding back a large flood. In the late 1970s the Cave Creek Dam was superseded by the U.S. Army Corps of Engineers' Cave Butte Dam, built a short distance downstream. The 1923 dam remains in place but is no longer required to provide flood control. Its use of a curved upstream face makes it unique among multiple-arch dams in the United States, and it stands as one of America's most remarkable reinforced-concrete structures because it uses so little material. The entire structure consists of about 19,000 cubic yards of concrete; thus, every linear foot of the 1,700-foot-long design required only slightly more than 10 cubic yards of concrete. Given that the structure averages at least 70 feet in height, it would be difficult to come up with a better example of material conservancy in dam design. Recently, David Billington, an engineering professor at Princeton, developed the notion of "structural art" that is based upon the ideals of efficiency, economy and elegance. In accordance with this definition, the Cave Creek Dam ranks as a premier example of structural art in the United States.

■ **Cave Creek Dam**
Across Cave Creek
1 mile west of Cave Creek
Road, 8 miles south of
Cave Creek
John S. Eastwood
1923

Curved upstream face of the Cave Creek Dam shortly after its completion in early 1923.

COOLIDGE DAM

Because the U.S. Bureau of Reclamation is the major federal agency involved in developing western water resources, it largely overshadows other federal activity in the field. It is little known that the Indian Irrigation Service began operation under the U.S. Department of the Interior in the early 20th century and made significant contributions toward increasing Native American agriculture in the West. In the 1920s the U.S. Indian Service (later called the Bureau of Indian Affairs) continued this work and undertook construction of the Coolidge Dam in south-central Arizona. Built to supply

■ **Coolidge Dam**
Across the Gila River
On Indian Route 3 in the
San Carlos Indian
Reservation
C. R. Olberg
1928

Left: Upstream face of the Coolidge Dam, showing the egglike shapes of the three domes. Right: Downstream face, with the spillway on the right.

water for the San Carlos Indian Reservation, the Coolidge Dam is a reinforced-concrete, multiple-dome dam with a maximum height of 249 feet. Similar to a multiple-arch buttress design, the multiple-dome design features an upstream face formed by domes that are curved around two axes. In contrast, cylindrical arches are curved around only a single axis. In structural terms, the dome develops strength in a manner analogous to the way an egg resists certain types of crushing. After World War II several very thin arch dams were built using "double-curved" designs, but the Coolidge Dam is the only multiple-dome dam in the United States. It has provided good service to the San Carlos Indians, but the flow of the Gila River is often insufficient to make full use of the dam's storage capacity. In fact, the famed humorist Will Rogers is reported to have quipped on visiting the newly completed dam, "If this was my reservoir, I'd mow it." The dam remains in active use and is a major example of federal support for Indian agricultural development. NR.

FLAGSTAFF

■ **Canyon Padre Bridge**
Across Canyon Padre
On the old U.S. Route 66
right-of-way, 22 miles east
of Flagstaff
Daniel Luten and Topeka
Bridge and Iron Company
1914

With the growth of automobile travel in the early 20th century, interest developed in building a highway between Santa Fe, N.M., and southern California. Originally called the Old National Trail Highway, this road later became part of the world-famous Route 66 stretching from Chicago to Los Angeles. While the climate of northern Arizona is quite dry, the highway still had to cross numerous creek beds *(arroyos)*. One of these was Canyon Padre, east of Flagstaff. In 1913 the newly formed Arizona State Highway Department requested competi-

the canyon. Shortly
...d Iron Company of
...epresentative of Daniel
...mpany, received a $7,900
...ot-long, single-span, concrete
...rom use on Route 66 in 1937, the
... carries local traffic on the nearby
...eservation. As documented by Clay
...cent state bridge inventory, the Canyon
...ge is among the oldest surviving reinforced-
...ete bridges in Arizona.

LAKE HAVASU CITY

The most incongruous historic bridge in the United
States is the five-span masonry and reinforced-concrete
structure located in the recently built resort community
of Lake Havasu City in western Arizona. Designed by the
famed British engineer John Rennie (and built by his son
of the same name), the bridge originally spanned the
Thames in London, England. Greeted with much fan-
fare, it first opened in 1831 to replace the "original"
London Bridge (1209) designed by Peter de Colechurch.
Rennie's 930-foot-long masonry arch bridge carried
highway traffic for more than 135 years and even survived
the German blitz of World War II, until transportation
officials in 1967 called for its replacement. Choosing not
to disperse or discard the facing stone from the 1831
bridge, the city of London sold 10,000 tons of it to the
McCulluch Oil Corporation for $2.46 million. In 1968
work on transferring the facing material from London to
Arizona began. After traveling by ship and rail, the
carefully marked pieces were reassembled on the surface
of a five-span reinforced concrete arch bridge designed
especially to mimic Rennie's original structure. The
result is a structurally bogus, but visually correct,
London Bridge. The frivolity of the setting, where
motorized aquatic sports (such as jet-ski racing) abound,
will no doubt offend serious bridge historians. But
Arizona's little bit of 19th-century England is nonetheless
an intriguing icon with a unique place in America's bridge
heritage.

■ **London Bridge**
Across an inlet of Lake
Havasu (Colorado River)
East of State Route 95
John Rennie
1831, 1971

MARBLE CANYON

The Navajo Bridge is the only highway bridge across the
Colorado River wholly within the state of Arizona.
Located near the site of the historic Lee's Ferry, a river
crossing first developed by Mormon settlers traveling
from Utah to Arizona, the bridge also allowed auto
tourists for the first time to visit both sides of the Grand
Canyon without traveling a circuitous route through
Arizona, California and Nevada. The structure is a steel
deck arch with a main span of 616 feet and two approach
spans totaling slightly more than 200 feet. The main arch
is a three-hinged structure with a rise of 103 feet, and the
bridge deck is 467 feet above the normal water level in the
Colorado River. Because of the impossibility of erecting

■ **Navajo Arch Bridge**
Across the Colorado River
On Alternate U.S. Route 89
R. A. Hanson
1929

Graceful Navajo Bridge spanning the formidable chasm cut by the Colorado River.

temporary falsework to support the bridge during construction, the structure was built by cantilevering out from both sides of the canyon. Financing for the bridge came from both the state of Arizona and the Navajo Tribal Fund. Since its completion, the bridge has remained in constant use and has not undergone any major alterations. The state department of transportation is planning to build a parallel structure to carry the highway, allowing the earlier bridge to be developed as a tourist attraction. NR.

MESA

■ **Granite Reef Dam**
Across the Salt River
5 miles east of State
Route 87
U.S. Reclamation Service
1908

Left: Granite Reef Dam during a period of low water. Right: Arizona Canal carrying water toward Phoenix.

Water stored behind Arizona's Roosevelt Dam (1911) is used for irrigation in the Salt River Valley. After release from Roosevelt Lake, this water flows down the Salt River, passes through hydroelectric plants at the Horse Mesa, Mormon Flat and Stewart Mountain dams and, about 55 miles downstream from the Roosevelt Dam, reaches the Granite Reef Dam. Here, water is diverted for use in Phoenix, Tempe, Mesa and other parts of the Salt River Valley. One of many western dams built by the Reclamation Service (forerunner of the Bureau of Reclamation), this structure is a 40-foot-high, 1,000-foot-long, concrete overflow dam that diverts water into the Arizona Canal on the north shore of the river and the Grand Canal on the south shore of the river. The present concrete dam was built to replace a timber crib dam built in the 1890s

that had suffered periodic damage from heavy flooding. Although not a tall, impressive-looking storage structure, the Granite Reef Dam is a critically important part of the Salt River Valley's water supply system.

PAGE

When plans for the first major storage dam across the Colorado River were being formulated in the early 1920s, many engineers and landowners in Arizona advocated construction at a site upstream from the Grand Canyon in the vicinity of Marble Canyon. At the same time, southern California real estate and agricultural interests favored building a dam in Nevada's Boulder Canyon, much farther downstream. The influence of the latter group prevailed, with the result that the Hoover Dam (1935) became the first storage dam completed across the Colorado. Ultimately, the plans for a dam in Marble Canyon led to construction in the early 1960s of the Glen Canyon Dam, designed solely to generate hydroelectric power. Located a few miles below the Utah-Arizona border, the 702-foot-high concrete arch dam is the tallest of its type in the United States. The dam impounds Lake Powell, a 186-mile-long reservoir that is a popular recreation site in the Southwest.

■ **Glen Canyon Dam**
Across the Colorado River
Adjacent to U.S. Route 89
U.S. Bureau of Reclamation
1963

PARKER

One of the primary purposes of the Hoover Dam (1935) near Boulder City, Nev., was to provide water for the rapidly growing region around Los Angeles. Community leaders from Los Angeles, as well as Pasadena, Anaheim and other communities, formed the Metropolitan Water District of Southern California in the late 1920s to facilitate construction of the Colorado Aqueduct. This aqueduct takes water out of the Colorado River and pumps it through a lengthy system of tunnels and pipelines to reach consumers in southern California. Water first passes through turbine-generator units at the Hoover Dam and then flows about 100 miles downstream to the Parker Dam. This 320-foot-high concrete arch structure impounds a portion of the river's flow and provides an intake reservoir for the aqueduct. The dam was built by the Bureau of Reclamation, but all construction financing came from the water district. The dam continues to serve southern California's water supply needs, but its reservoir now is also used as a source for the newly built Central Arizona Project that carries Colorado River water to the greater Phoenix-Tucson area.

■ **Parker Dam**
Across the Colorado River
East of State Route 95,
15 miles north of Parker
U.S. Bureau of Reclamation
1938

ROOSEVELT

When Anglo-Americans settled in Phoenix in the late 1860s, their first irrigation canals were built along the lines of water ditches dug by the Hohokam Indians hundreds of years before. Like the Hohokam, these settlers drew water from the Salt River as it flowed westward toward the Colorado River and the Gulf of

■ **Roosevelt Dam**
Across the Salt River
Adjacent to State Route 88
(The Apache Trail)
U.S. Reclamation Service
1911

California. During the late 19th century irrigation in the Phoenix area continued to grow but without the benefit of a large dam that could store spring floods for use during the rest of the year. A site about 60 miles east of, or upstream from, Phoenix near the junction of Tonto Creek and the Salt River appeared to be an ideal spot for a dam. The difficulty, however, lay in finding a way to finance construction of a large water impoundment structure in such a remote location.

The solution came with the establishment of the Reclamation Service in 1902. After a bit of political maneuvering, the Reclamation Service soon announced that one of its first projects would be to build a large storage dam at the Tonto site. Water from the dam's reservoir would be released for use by the Salt River Valley Water Users Association, an organization consisting of most major irrigation interests in the Phoenix area. Ironically, although the Reclamation Service had been specifically charged with building projects that would open up the West to new settlers, all of the irrigable land in the project area was already in private hands. Unfortunately, this was not the first, or last, time that rhetoric has not matched reality in the arid West.

To provide construction access, the Reclamation Service first had to build a road from Apache Junction to the dam site. Called the Apache Trail, the road was completed in March 1905, and, today, it passes through some of Arizona'a most spectacular desert scenery. Construction of the project's 280-foot-high, curved gravity masonry dam began in 1906 and, after several delays and cost overruns, was completed five years later. The dam is named in honor of President Theodore Roosevelt. In fact, Roosevelt, a major supporter of the Reclamation Service, was on hand for the official opening of the dam in 1911. Although the Roosevelt Dam cost three times more than its original estimate, it soon proved a technological success in increasing Phoenix's water supply. More than anything else, it allowed for the city's remarkable growth in the 20th century. Now plans have been devised to raise the structure about 70 feet to increase its storage capacity as part of the federally sponsored Central Arizona Project. These plans will entail encasing most of the 1911 structure in concrete, covering up the original masonry in all its splendor. NR, ASCE.

Left: Roosevelt Dam across the Salt River. The hydroelectric plant is in the foreground and Lake Roosevelt in the background. Right: View of the dam in 1941, highlighting the pronounced curve of the design.

Topock Bridge after being converted to carry an oil pipeline.

TOPOCK

The Topock Bridge is a 592-foot-span steel through arch built to carry highway traffic across the Colorado River between Arizona and California. Originally a component in the Old National Trail Highway, it later became part of the famed U.S. Route 66. In the late 1940s the Santa Fe Railway replaced its nearby Red Rock Cantilever Bridge (1890) with the Topock Bridge and donated the Red Rock span (since demolished) to Arizona and California for highway use. In 1948 the arch bridge was sold to the El Paso Natural Gas Company and the Pacific Gas and Electric Company. The firms subsequently modified it to carry a gas pipeline over the river. Although no longer serving its original purpose, the main structure of the Topock Bridge retains its historical integrity. It is readily visible from the modern highway bridge that carries Interstate 40 across the Colorado River.

■ Topock Bridge
Across the Colorado River
Adjacent to Interstate 40
J. A. Sourwine
1916

YUMA

Among the first projects built by the Reclamation Service was the 4,800-foot-long rockfill Laguna Dam across the lower section of the Colorado River. The structure is part of the Yuma Irrigation District and is used to divert water for agricultural development in Arizona and California. Because of the powerful flow of the Colorado River, which can exceed 100,000 cubic feet of water per second,

■ Laguna Dam
Across the Colorado River
6 miles north of U.S.
Route 95
U.S. Reclamation Service
1909

Headgates of the canal at the Laguna Dam, which irrigates lands near Yuma, Ariz.

the dam took four years to complete. Although only 19 feet high, the rockfill structure has a width of more than 225 feet and required a huge amount of material to construct. After 75 years of operation, the Laguna Dam still diverts water for irrigation use in the Yuma area.

■ ■ ■ ■ ■ ■   NEW MEXICO   ■ ■ ■ ■ ■ ■

CONCHAS

■ **Conchas Dam**
Across the Canadian River
Adjacent to State Route 129
U.S. Army Corps of
Engineers
1938

During the early 20th century the Army Corps of Engineers played no role in the development of water resources in the arid West. Such activity was left to private organizations, state agencies, municipalities and the U.S. Reclamation Service. With the economic devastation of the 1930s, the Roosevelt administration found a use for the corps in the region. Dam construction projects offered a way to pour money into the regional economy, put people to work and provide such long-term benefits as the increased availability of water. In the mid-1930s New Mexico officials pushed hard for federal financing of a large storage dam on the upper Canadian River in the northeastern part of the state. The structure was intended to provide flood control for the town of Tucumcari and facilitate the growth of irrigated agriculture. After considerable political horse trading among government officials and large landholders in the region, the Conchas Dam received final approval in November 1935.

Conchas Dam, a concrete gravity design that was the Army Corps of Engineers' first major project in New Mexico.

Work on the 200-foot-high, 1,250-foot-long, concrete and earthfill gravity dam began shortly afterward. During the three years of construction, more than 3,000 men worked at the dam site and the project became an important source of income for the region's ravaged economy. The Conchas Dam began impounding water in December 1938, and it continues in operation. Since the dam's completion, the Corps of Engineers has become involved in numerous other water projects in the state.

EAGLE NEST

The Cimarron River flows east out of the Sangre de Cristo Mountains in northeastern New Mexico and passes into the panhandle of western Oklahoma. In the 1880s and 1890s, small irrigation projects were developed along the waterway, but none involved construction of a large-scale storage reservoir. In the late 19th century it became apparent that a tall dam built in the Moreno Valley along the upper reaches of the river could foster irrigation in a huge area downstream. Another favorable aspect of this site, which was more than 7,000 feet high, was that its relatively high elevation would limit the amount of evaporation loss from the reservoir. In 1906 Charles Springer and several associates bought the dam and reservoir site and, acting under the name Cimarron Valley Land Company, set out to find financing to construct the dam. Initially, the search proved fruitless, and the company developed a few small diversion dams to make some of its land productive. But in 1916 Springer chose to proceed with the large dam using his own money, and by 1918 construction was under way.

Because of his interest in minimizing the structure's cost and because of the site's hard granite foundations, Springer chose to build a 120-foot-high concrete arch dam with a maximum thickness of 52 feet. Today, the Eagle Nest Dam stands as testimony to the role of private capital in financing development of western water resources. The dam continues to be a major contributor to the economy of Colfax County, and its reservoir is a popular recreation site. NR.

■ **Eagle Nest Dam**
Across the Cimarron River
Near U.S. Route 64
Willis Ranney
1920

ELEPHANT BUTTE

Rising out of the Sangre de Cristo Mountains in south-central Colorado, the Rio Grande passes through the entire state of New Mexico before forming the international border between Texas and Mexico. Until the Elephant Butte Dam was built by the Reclamation Service, no large storage dams were located anywhere along the river. Impetus for constructing the massive 306-foot-high, 1,674-foot-long, concrete gravity dam came from irrigation interests in southern New Mexico and western Texas and from a 1907 treaty with Mexico stipulating that 60,000 acre-feet of water would be

■ **Elephant Butte Dam**
Across the Rio Grande
Adjacent to State Route 52
U.S. Reclamation Service
1916

Massive Elephant Butte Dam, with its vertical upstream face.

provided to Mexico from a dam on the Rio Grande.

Work on the Elephant Butte Dam began in 1911, and the first concrete was poured in June 1913. The dam required a construction camp with quarters for more than 1,000 workers; during its short life, it was the largest town in the county. The final section of concrete was placed in January 1916, and the dam immediately began impounding water. Its reservoir has since become a major recreation attraction in the region, and the dam continues to help irrigate several thousand acres of land in New Mexico and Texas. NR, ASCE.

FORT SUMNER

**■ Fort Sumner
Railroad Bridge**
Across the Pecos River
On the Santa Fe Railway
right-of-way, ½ mile north
of U.S. Route 60
American Bridge Company
1906

The steel-plate girder bridge is the unappreciated step-child of historic bridges. This structural form has been used since the mid-19th century for railroad structures and supports innumerable highway spans built in the early to mid-20th century. But, because they are so common and their bulky dimensions do not strike most people as aesthetically pleasing, they rarely attract historical attention. An exception is the Fort Sumner Railroad Bridge in eastern New Mexico. A 15-span, 1,500-foot-long, steel-plate girder bridge supported on concrete piers, it was built by the Atchison, Tokepa and Santa Fe Railway as a primary structure on the Belen Cutoff between Belen, in central New Mexico, and Texico, on the New Mexico–Texas border. After more than 80 years, the bridge remains in constant use. NR.

Multispan steel girder bridge at Fort Sumner, supported on concrete piers.

LAKEWOOD

■ McMillan Dam
Across the Pecos River
5 miles east of U.S.
Route 285
Louis Blauvelt
1893, 1937

Built by private capital, the McMillan Dam constituted a key component of one of the West's largest 19th-century irrigation projects. Constructed to an initial height of 52 feet, the 1,687-foot-long rockfill and earthen structure was located in a relatively porous part of the Pecos River Valley. As a result, the owners experienced problems with reservoir water seeping into the foundations. The irrigation project teetered on the brink of insolvency, but, after the exertion of considerable political pressure by territorial officials in 1907, the U.S. Reclamation Service soon took over and refurbished the structure. In 1937 the Bureau of Reclamation increased the dam's height to 57 feet to compensate partially for the loss of reservoir space resulting from silt accumulation. Plans are under consideration for a further increase in Lake McMillan's capacity, which will only enhance the reservoir's significance in the regional economy.

OTOWI

After New Mexico gained statehood in 1912, the problem of improving the territorial road system became a subject of great importance. The relatively few taxpayers in the state could not afford to finance an elaborate, expensive highway system, and means were sought to reduce capital expenditures wherever possible. Given the great distances between towns, this was not an easy task. Bridges posed a special problem because the state's usually dry rivers would periodically flood and wash away any small highway spans. Between 1910 and 1930 the New Mexico State Highway Department built several suspension bridges that allowed placement of the support towers above normal flood levels but did not require huge amounts of money to construct. A large and significant example of this type of structure is the department's 174-foot-long Otowi Suspension Bridge across the Rio Grande, northwest of Santa Fe. Completed in 1924, it consists of reinforced-concrete towers, steel wire cables and a combination wood and metal Howe truss to support the roadway. The bridge became an important crossing of the Rio Grande and was used to help transport equipment during the initial establishment of the Los Alamos Scientific Laboratory as part of the famous Manhattan Project. The Otowi bridge was removed from highway service in 1948; for many years afterward it found use as a crossing for sheep and other livestock but today it is closed to all traffic. NR.

■ **Otowi Suspension Bridge**
Across the Rio Grande
On State Route 4
James A. French
1924

Otowi Bridge, with its wire cables supported on concrete towers.

■ ■ ■ ■ ■   OKLAHOMA   ■ ■ ■ ■ ■

LANGLEY

**■ Grand River
(Pensacola) Dam**
Across the Grand River
On State Route 28
Victor Cochrane
1940

In 1935 the Oklahoma legislature authorized creation of the Grand River Dam Authority to provide hydroelectric power and flood control for the northeastern part of the state. After weathering challenges to its legality, the authority began construction in 1938 of a multiple-arch dam across the Grand River. Completed in 1940, the Grand River Dam is the longest and last major multiple-arch dam built in the United States. It consists of 51 arches, each 84 feet wide, that form the upstream face of the dam. The maximum vertical height of the arches is almost 150 feet. In addition to a 4,300-foot-long, multiple-arch section, the dam includes an 861-foot-long concrete gravity spillway. The dam forms the Lake of the Cherokees, a popular recreation site in Oklahoma. It is easily accessible to visitors as State Route 28 runs across the top of the dam.

Left: Grand River Dam.
Right: Multiple-arch
upstream face of the dam.

 he Beautiful Grand River Dam.

OLUSTEE

■ Fullerton Dam
Across Turkey Creek
6 miles north of State
Route 44
J. William Fullerton, builder
c. 1895

It is easy to look back at the history of America's engineering development and focus only on the successes that, in retrospect, appear almost inevitable. But the evolution of technology also includes failed endeavors, and this is something that should not be ignored. In the late 19th and early 20th centuries, irrigation projects were undertaken by a wide variety of groups and individuals with no guarantee of success. Ultimately, the only way to ascertain the economic feasibility of a project was to go out and try to build it. If it succeeded, the promoter was hailed as a visionary and a sagacious businessperson. If it failed, the promoter could only move on to another endeavor and hope for better luck in the future.

The Fullerton Dam in southwestern Oklahoma is an example of an irrigation project that initially succeeded but eventually foundered and left its owner in financial disarray. Built by J. William Fullerton around 1895, the 25-foot-high masonry gravity dam diverted water from Turkey Creek for use by farmers in the Olustee area. Fullerton's operation proved profitable for several years, but by 1910 falling agricultural prices caused him to go into debt to keep the enterprise functioning. When he died in 1906, his holdings were sold to satisfy debts against his estate. New owners were initially interested in

Above: Fullerton Dam, with its builder, J. William Fullerton, standing on the crest, c. 1900. Left: Remains of the dam today.

further developing the Fullerton Dam operation, but, after a flood washed away much of the structure in 1919, the facility was permanently abandoned. Today, the masonry remains of the Fullerton Dam are readily visible and stand as mute testament to the thousands of pioneers who worked hard to develop western America but received little material reward for their effort. They, too, are important in our nation's history. NR.

■ ■ ■ ■ ■ ■ ■ TEXAS ■ ■ ■ ■ ■ ■ ■

BURNET

Although not as large as the more famous Colorado River that flows through the Grand Canyon in Arizona, the Colorado River in Texas is one of the largest rivers in the state. In the early 1930s Samuel Insull, the renowned Chicago utility magnate, became involved in a project to develop hydroelectric power along the river in the hill country northwest of Austin. The Insull-controlled Central Texas Hydro-Electric Company soon began work on the George W. Hamilton Dam (named after one of Insull's engineers), a large multiple-arch structure with a maximum height of 145 feet above bedrock. However, Insull's financial empire soon collapsed in the midst of the Depression, and the partially completed dam lay abandoned from 1932 to 1934.

In 1934 the Texas legislature created the Lower Colorado River Authority to take over the Hamilton Dam

■ Buchanan Dam
Across the Colorado River
Adjacent to State Route 27,
10 miles west of Burnet
Fargo Engineering Company,
with B. F. Jakobsen and
Victor Cochrane
1938

Incomplete Buchanan Dam in 1932, after the collapse of Samuel Insull's financial empire. The Lower Colorado River authority took over, finishing construction in 1938.

and, with funding from the federal Public Works Administration, efforts to complete the structure soon began. At this time it was renamed the Buchanan Dam in honor of Rep. James P. Buchanan, who pushed for federal support for the project. The dam is more than two miles long with a main multiple-arch section more than 2,000 feet in length. Located at the deepest part of the dam, this section consists of 29 arches, each with a span of 70 feet. The dam forms a reservoir with a capacity of 992,000 acre-feet of water. After 50 years of operation, the Buchanan Dam is still among the most important facilities operated by the Lower Colorado River Authority.

CISCO

■ **Williamson Dam**
Across Sandy Creek
Adjacent to State Route 6,
6 miles north of Cisco
Henry Exall Elrod
Engineering Company
1924

The Williamson Dam is an excellent example of a reinforced-concrete, flat-slab buttress typical of those first popularized in the early 20th century by the Ambursen Hydraulic Construction Company of Boston. To expand its water supply, Cisco chose in the early 1920s to build a storage dam on nearby Sandy Creek. After construction of a spur line from the Missouri, Kansas and Texas Railroad to deliver equipment and supplies to the site, work on the dam took more than two years to complete. The structure is 1,060 feet long and has a maximum height of 133 1/2 feet; the overflow spillway is placed in the middle of the dam. A roadway extends along the dam's crest supported by buttresses. The Williamson Dam remains today an integral part of Cisco's water supply system.

Williamson Dam, with its overflow spillway and highway across the top.

Left: Scherzer rolling lift bridge, which, as part of the Galveston Causeway, provided clearance for ship passage. Below: Train on the causeway's concrete arch structure.

GALVESTON

Beginning in the 1850s the first of three wooden railway trestles was built to connect the Texas mainland with the port of Galveston. In the 1890s a steel-truss highway bridge was also built across Galveston Bay. All four bridges spanning the bay, however, were destroyed during the hurricane of 1900. After the repair of one of the railway trestles, civic attention soon turned to the problem of building a permanent highway-railroad bridge across the bay. After protracted negotiations with numerous railroads, work on the causeway began in 1909. Completed in 1912, this structure consisted of more than 8,000 feet of earthen embankment and 28 reinforced-concrete arches, each with a span of 70 feet. In addition, in the center of the reinforced-concrete section was a Scherzer rolling lift bridge designed to allow passage of large ships.

In 1915 a violent hurricane severely damaged the earthen embankments, but the reinforced-concrete arches survived intact. During the next seven years the earthen sections were replaced with reinforced-concrete arches similar to those in the original design. The refurbished Galveston Causeway operated without incident for many years until it was recently rendered obsolete by a completely new bridge. Today, it is closed to vehicular traffic but can be used by hardy pedestrians who want to walk between Galveston and the mainland.

■ **Galveston Causeway**
Across Galveston Bay
Adjacent to Interstate 45
Concrete Steel Engineering Company
1912, 1922

MARSHALL FORD

After its founding in 1902, the U.S. Reclamation Service (later the Bureau of Reclamation) built dams in almost all the arid western states. The major exception was Texas, where, because of agreements made when it became part of the United States in 1845, no public land was owned by the federal government. It was not until the mid-1930s that the Lower Colorado River Authority sought to bring a Bureau of Reclamation–financed project into the state. The resulting Marshall Ford Dam, soon renamed in honor of Texas Rep. Joseph J. Mansfield, used a fairly standard concrete gravity design. But it is historically noteworthy because one of Lyndon B. Johnson's first accomplishments as a freshman con-

■ **Mansfield (Marshall Ford) Dam**
Across the Colorado River
Adjacent to State Route 620
U.S. Bureau of Reclamation
1942

Aerial view of the Mansfield Dam and Lake Travis, c. 1945. Construction of this dam played a major role in Lyndon Johnson's early political career.

gressman was passage of special legislation authorizing construction of this Bureau of Reclamation dam on nonfederal land. As documented in Robert Caro's *The Path to Power*, Johnson's effort was especially critical for the contractors, who were already deeply involved in construction when final federal approval came through. Today, the 278-foot-high, 5,093-foot-long Mansfield Dam remains an important part of the Lower Colorado River Authority's regional water control system.

SAN ANTONIO

■ Espada (San Antonio) Aqueduct Bridge
Across Piedras Creek
Along the Espada Acequia
c. 1740

While British and other settlers were struggling to establish colonies on the eastern coast of North America in the 17th and 18th centuries, Spanish settlers from Mexico were moving north into areas that later became part of the southwestern United States. One of the largest of these Spanish settlements centered around the San Antonio River in south-central Texas. Beginning with construction of the Alamo Madre Acequia in 1718, designed to serve lands surrounding the San Antonio de Valero Mission, several irrigation canals, often called *acequias* (ditches), were built in the San Antonio region. These irrigation systems diverted water from the San Antonio River using small overflow dams. The water was then carried to arable fields in short canals only a few miles in length. Most of the canals were built by Catholic missions, including the Espada Acequia, which is now the oldest active irrigation system in the United States. This acequia, which is four and one-half miles long and can deliver approximately 12 cubic feet of water per second, is especially noteworthy because it includes a two-span stone aqueduct that carries the canal across Piedras Creek. This structure, with arches that are 12 feet and 16½ feet in length, stands as one of the oldest surviving bridges in the United States. ASCE.

TYLER

■ Tyler Dam
Across Indian Creek
On Bellwood Road,
½ mile south of State Route 31
Julius M. Howells
1894

Although a few earlier earthfill dams were built in part using the hydraulic fill process, the Tyler Dam is considered the first built entirely by this method. Hydraulic fill technology developed from California gold-mining practices in which pressurized water was directed against hillsides or other large earthen mounds. Miners would then process the resulting muck to extract gold. Beyond its specific role in gold mining, hydraulic fill technology is an efficient way to transport large quantities of earth at relatively little cost. It was in this latter context that Julius M. Howells devised a means of building up earth dams using hydraulic fill technology.

Work on the project for the Tyler Water Company started in May 1894 when Howells built a water-pumping plant near Indian Creek about five miles west of the center of town. Water was pumped to a height about 75 feet above the proposed dam site and directed by firehose onto an adjacent hill. The resulting muck flowed downhill and was carried through wooden flumes to various parts of the dam site. The soggy earthfill was then allowed to drain away its excess water before being used to impound a reservoir for the community water supply. The Tyler Dam is 575 feet long with a maximum height of 32 feet. Its reservoir maintains a maximum capacity of 1,770 acre-feet (or more than 500 million gallons). The dam is no longer a component of Tyler's water supply system, but its reservoir (Lake Bellwood) is the focus of a regional park and picnic area. NR.

WACO

During the Reconstruction era following the Civil War, Texas struggled to reestablish an economy ravaged by "the late unpleasantness." With the beginning of the Chisholm Trail cattle drives around 1867, the Brazos River Valley entered a new era of relative prosperity. At the same time, Waco businesspeople sought a permanent means of overcoming the commercial barrier formed by the often-treacherous Brazos River. In 1866 the state legislature chartered the Waco Bridge Company to build a crossing for the stream and authorized it to collect tolls. The company chose to build a suspension span and in 1868 began work on a 475-foot-long bridge designed in concert with John A. Roebling (1806–69), the engineer of the Brooklyn Bridge (1883) and other notable spans. The Waco Bridge used brick manufactured in Texas for the castlelike support towers, but iron for the cables and traffic deck came from Roebling's New Jersey–based firm. Texas engineering historian T. Lindsay Baker has noted that "the roadway was so wide that two stagecoaches could pass each other going in opposite directions" and that "no other bridge in the state for years could compare with it in either scale or beauty."

The bridge came under the control of the city of Waco in 1889 and became a "free" bridge. In 1913–14 new steel cables and a new steel deck structure replaced the original components supplied by Roebling. In addition, the masonry towers were modified to provide a simpler, less cluttered appearance. The Waco Suspension Bridge was removed from active use in 1971 and is now maintained by the city as a pedestrian walkway and historic monument.

■ **Waco Suspension Bridge**
Across the Brazos River
At Bridge Street
Thomas M. Griffith and
John A. Roebling and Son
(original)
G. E. Byars,
R. J. Winslow and Missouri
Valley Bridge and Iron
Company (reconstruction)
1869, 1914

Water cascading over the spillway of the Grand Coulee Dam (1942) in eastern Washington. The reservoir behind the 550-foot-high dam stretches upstream for more than 150 miles.

CHITINA

■ Kuskalana Bridge
Across the Kuskalana River
At Mile 146 on the Copper
River and Northwestern
Railroad
A. C. O'Neel
1910

Located in a remote part of southeastern Alaska, the Kuskalana Bridge and the railroad it carried were built to provide a means of transporting the rich copper deposits of the region to the Pacific Coast. Both the bridge and the Copper River and Northwestern Railroad line served the Kennecott copper-mining operations. The bridge is a three-span, pin-connected, steel Pratt deck truss with wooden trestle approach spans. The overall length of the steel trusses is 545 feet, and the bridge's maximum height is 240 feet above the river. No longer carrying rail traffic (the Kennecott mine has not operated since World War II), the bridge is used only infrequently by hardy travelers with four-wheel-drive vehicles.

Kuskalana Bridge. Right: All-steel pier. Below: Pratt deck truss, which carried trains laden with copper toward the southern Alaskan coastline.

JUNEAU

In the 1880s gold was discovered in Silver Bow Basin near Juneau, the territorial capital of Alaska. The Alaska-Juneau Mining Company operated as one of the largest gold producers in the region, and in the early 20th century it acted to increase its production capacity. An important part of this expansion involved construction of a large concrete dam to supply water for the company's two hydroelectric plants on Salmon Creek. To reduce the amount of concrete required (and thus limit the dam's cost), the company selected a 168-foot-high, 640-foot-long, constant-angle arch design prepared by Lars Jorgensen, a Swiss-born engineer who had moved to San Francisco to work in hydroelectric power engineering. By using a constant angle at every elevation, it was possible to reduce the concrete required without increasing the stresses in the arch. Although other engineers had previously discussed the concept of building constant-angle arch dams, Jorgensen was the first to design and build such a structure. The Salmon Creek Dam is the first constant-angle arch dam built in the world. After almost 75 years of service, it still functions as part of the Alaska Electric Light and Power Company's regional electric power system.

■ **Salmon Creek Dam**
Across Salmon Creek
6 miles northwest of Juneau
Lars Jorgensen
1914

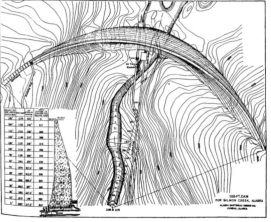

Above: Crest of the Salmon Creek Dam. Left: Drawing highlighting the conical shape of the constant-angle design.

■ ■ ■ ■ ■ ■ ■ CALIFORNIA ■ ■ ■ ■ ■ ■ ■

BIXBY CREEK

■ **Bixby Creek Bridge**
Across Bixby Creek
On State Route 1
F. W. Panhorst and
C. H. Purcell
1933

In designing a highway south from Carmel along the dramatically beautiful Pacific coastline, the California Division of Highways sought to complement nature with aesthetically pleasing structures of the division's own. The most famous of these is the 714-foot-long, reinforced-concrete, open-spandrel arch bridge over Bixby Creek. With a main arch span of 330 feet and a clearance of more than 280 feet between the creek bed and the top of the bridge, it is among the lightest and most graceful structures of its type in the United States. The design's only incongruous features are the heavy piers located directly over the arch abutments. However, these piers do not negate the bridge's overall aesthetic power. The bridge remains in service as part of California's state highway system.

BRIDGEVILLE

■ **Lower Blackburn Grade Bridge**
Across the Van Duzen River
On State Route 36
John B. Leonard
1925

In the early 1920s the California Highway Commission selected the Fortune–Red Bluff Highway, later State Route 36, as the most desirable route between the state's northern coast and the upper Sacramento Valley. Humboldt County soon began work on improving this route and contracted for five new bridges along the right-of-way. John B. Leonard (1864–1945), a prominent California engineer who specialized in reinforced-concrete designs, received the commission for this work. As documented by John Snyder, a historian with the California Department of Transportation, Leonard played a critical role in rebuilding San Francisco after the 1906 earthquake and also served for many years as an associate editor of the influential journal *Architect and Engineer of California*. During his career he designed scores of bridges and buildings, and the Humboldt County project was one of his last major commissions. The Lower Blackburn Grade Bridge is a 258-foot-long rainbow arch design and is perhaps the most visually attractive of Leonard's surviving bridges. The structure was recently removed from highway use and is now accessible only to pedestrians. NR.

CANYON DAM

■ **Big Meadows (Lake Almanor) Dam**
Across the North Fork of the Feather River
On State Route 89
Julius M. Howells
1913, 1927

In 1902 Julius M. Howells filed claims to build a large dam on the North Fork of the Feather River to generate hydroelectric power. The plan involved storing water at Big Meadows, a large valley more than 4,000 feet above sea level, and then releasing this water through a series of hydroelectric plants before it reached the Sacramento River near Oroville. By 1911 the Great Western Power Company was ready to build the Big Meadows Dam, and construction began on a 150-foot-high, reinforced-concrete, multiple-arch structure. However, following a change in corporate management (and considerable

Above: Bixby Creek Bridge amid the rugged terrain of the Big Sur. Left: John Leonard's rainbow arch design near Bridgeville. Below: Remains of the aborted multiple-arch dam at Big Meadows, 1918. The spillway for the earthfill replacement dam is visible on the left.

lobbying on the part of John R. Freeman, a prominent eastern engineer opposed to multiple-arch dams), the company altered its plans and built a 110-foot-high, hydraulic fill earthen dam. Although this structure proved more expensive than the multiple-arch design, its massive appearance apparently made the company feel more secure in building it. In 1927 its height was raised to about 150 feet above the foundations. Today, the 1,250-foot-long dam is the key component of the Pacific Gas and Electric Company's Feather River power-generating system. The reservoir is known as Lake Almanor, and it is a popular recreation site in the northern Sierra Nevada.

ELK CREEK

■ Stony Gorge Dam
Across Stony Creek
5 miles south of State
Route 162
Ambursen Hydraulic
Construction Company
1928

Compared with other western states, California received little benefit from U.S. Reclamation Service or Bureau of Reclamation projects before the 1930s. Although, in general, California's agricultural interests were distrustful of federal projects before the New Deal, the Orland Project in the northern part of the state was a major exception. This development project, centered approximately 125 miles north of San Francisco, provided storage for water flowing eastward from the Coast Mountain Range into the Sacramento Valley. In the late 1920s the Bureau of Reclamation built the 868-foot-long Stony Gorge Dam to supply additional storage capacity for irrigation on the western side of the valley. With buttresses spaced 18 feet apart and sloped at 45 degrees on the upstream face, the Stony Gorge Dam is the only major flat-slab buttress dam built by the bureau. Because of the use of a buttressed design, the bureau was able to save considerable quantities of material over a comparable concrete gravity design.

EMIGRANT GAP

■ Spaulding Dam
Across the South Fork of
the Yuba River
1 mile north of Interstate 80
James H. Wise, Frank F.
Baum and Lars Jorgensen
1913, 1916, 1919

In 1912 the Pacific Gas and Electric Company began work on a new concrete curved gravity dam to impound the headwaters of the South Fork of the Yuba River. Designed to replace an 1892 rockfill dam built by the South Yuba Water Company (a mining operation), the new dam was part of an elaborate hydroelectric power project that diverted water from the Yuba River watershed to power plants along the Bear River directly to the north. Construction started in June 1912, and a year later the structure had reached a height of about 65 feet. At that time Pacific Gas and Electric changed the design to a constant-angle arch dam that would be much thinner, and hence less expensive, than the original gravity design. Thus, the completed structure is a concrete arch dam that rests on a large gravity base. By the end of 1913 the dam was 225 feet high; in 1919 the dam reached its present height of 275 feet. Located directly adjacent to Interstate 80, it is among the largest and most accessible hydroelectric dams in California. It is named after John Spaulding, an important figure in the history of the South

Spaulding Dam as it looked in 1920.

Yuba Water Company. The Spaulding Dam remains in service as a key component of Pacific Gas and Electric's hydroelectric power system.

FRIANT

In the early part of the 20th century, the U.S. Reclamation Service, later the Bureau of Reclamation, undertook little work in California. The reasons are complex, but they relate primarily to the extensive amount of privately financed irrigation development in the state. These private interests initially felt threatened by federally sponsored projects. In the 1930s, however, interest in the Bureau of Reclamation entered a new phase when California sought federal support to build large dams that would dramatically increase the water supply in the state's most agriculturally productive region. Known as the Central Valley Project, its implementation involved the construction of two major dams, the Shasta Dam (1945) on the Sacramento River north of Redding and the Friant Dam on the San Joaquin River near Fresno. Water impounded by the Shasta Dam, a 319-foot-high, 3,488-foot-long, concrete gravity structure, is made available for irrigation on land surrounding the lower

■ **Friant Dam**
Across the San Joaquin River
Adjacent to State Route 145
U.S. Bureau of Reclamation
1942

Friant Dam under construction in 1941. The hulking gantry cranes were used to place concrete in the formwork.

Downstream side of the Friant Dam, with the San Joaquin River in the foreground.

sections of the river. In turn, this allows water from the upper San Joaquin River to be diverted at the Friant Dam and carried by canal to dry, fertile land in Kern County to the south. In essence, the Shasta and Friant dams work in tandem to facilitate a huge water swap that allows additional irrigation in the southern part of the Central Valley. The Friant Dam's role in increasing the region's agricultural productivity is undeniable. But it has also prompted criticism that federally subsidized water is being used primarily to foster large agribusiness development at the expense of smaller-scale family farms.

GRASS VALLEY

■ **Bridgeport Covered Bridge**
Across the South Fork of the Yuba River
Between State Route 49 and State Route 20
David Ingefield Wood, builder
1862

Following the spectacular gold rush into California during the late 1840s and early 1850s, prospectors and miners soon began exploring other parts of the West in search of new bonanzas. Among the richest new mining areas found was the Comstock Lode in western Nevada near Lake Tahoe. Centered around Virginia City, the Comstock Lode developed into one of the world's greatest sources of silver during the late 19th century. Although only about 150 miles from Sacramento in California's Central Valley, the Comstock Lode lay east of the

Auxiliary arch supporting the Bridgeport Covered Bridge, evident from the pattern of the exterior siding.

imposing barrier formed by the Sierra Nevada. To meet the transportation needs of the new mining region, private toll-road companies built highways to provide access to California and the trading ships that plied the Pacific Ocean. Among the most important of these was the Virginia City Turnpike Company. As part of its toll road aroes the South Fork of the Yuba River, the company built the 233-foot-long Bridgeport Covered Bridge, a Howe truss with an auxiliary arch to reduce deflection at the center of the span. It is the longest single-span covered bridge in the United States. Although only a handful of covered bridges survive in California, the technology played an important role in the state's transportation development during the 19th century. This span, which no longer carries vehicular traffic, is now maintained as part of the state park system. NR, ASCE.

HUME

In late 1905 the Michigan-based Hume-Bennett Lumber Company purchased the vast timber holdings of the bankrupt Moore-Smith interests in the central Sierra Nevada about 50 miles east of Fresno. This land contained many giant sequoias, the largest trees in the world, as well as numerous other big pines. Within a short time, the company recognized the need to build a new sawmill higher in the mountains surrounding the Kings River to harvest trees that could not be economically processed at the existing sawmill. Seeking to reduce its plant investment, the Hume-Bennett Company soon contracted with John S. Eastwood (1857–1924) of Fresno to build the world's first reinforced-concrete, multiple-arch dam at Long Meadow on Ten-Mile Creek. A resident of Fresno since 1883, Eastwood helped survey lumber flumes in the Sierra Nevada before devoting himself to hydroelectric power development in the mid-1890s. Between 1905 and 1907 he devised his multiple-arch dam design for the Big Creek hydroelectric

■ **Hume Lake Dam**
Across Ten-Mile Creek
In the Sequoia National
Forest, near State Route 180
John S. Eastwood
1909

Hume Lake Dam, with the reservoir in the background.

Reinforced-concrete buttresses supporting the Hume Lake multiple-arch dam.

power system on the San Joaquin River. However, he was rebuffed by the management of the Pacific Light and Power Company when he pushed for construction of the innovative design. To prove the viability of his new water-storage structure, he convinced the Hume-Bennett Company to let him design and build its new dam on Ten-Mile Creek.

The 650-foot-long Hume Lake Dam consists of 13 reinforced-concrete arches, each with a span of 50 feet. It has a maximum height of 61 feet and rests on solid granite foundations. The dam provided a storage pond for timber waiting to be processed in the adjacent sawmill and also supplied water for a 60-mile-long flume used to ship cut timber to the company's railhead at Sanger in the San Joaquin Valley. Lumbering in the region stopped in the 1920s, and in 1935 the U.S. Forest Service took control of the dam. The Hume Lake Dam now is used solely for recreational purposes, and the lake is a major tourist attraction in the Sequoia National Forest.

LITTLEROCK

■ Littlerock Dam
Across Little Rock Creek
Adjacent to Littlerock Dam
Road, 4 miles south of State
Route 138
John S. Eastwood
1924

At the time of its completion, the 175-foot-high Littlerock Dam was the tallest multiple-arch dam in the United States. The reinforced-concrete structure is more than 700 feet long and consists of 28 arches, each with a span of 24 feet, built with an upstream angle of 45 degrees. The original siphon spillways were damaged in the heavy floods of March 1938, which also caused the reservoir to overtop the dam by two feet, and these were replaced by the existing overflow spillway. Otherwise, the dam has undergone little alteration. The dam's origins date to 1892, when a small group of farmers in northern Los

Above: Detail of the Littlerock Dam's upstream face. Right: Aerial view showing the angle in the dam. The reservoir is on the left.

Angeles County formed the Littlerock Creek Irrigation District. Organized under the auspices of California's Wright Act, a law designed to spur irrigation in the state but branded by some as socialistic, the district struggled for many years because of water-shortage problems. In 1917 the district, along with the newly formed Palmdale Irrigation District, began serious planning for a storage dam on Little Rock Creek. The two districts possessed limited financial resources, and a multiple-arch dam was the only type of dam economically feasible for them to build. California's dam-safety bureaucracy required John S. Eastwood to prepare three different designs over a period of four years before finally approving construction in November 1922. The dam's safety remains a controversial subject with California's Division of Safety of Dams, although the structure has provided exemplary service without incident to the farmers of Littlerock for more than 60 years (see page 74). Completed only a few months before Eastwood's death in August 1924, the Littlerock Dam is the tallest of his 17 multiple-arch dams. It serves as an impressive monument to Eastwood's pioneering work in developing and promoting this important type of hydraulic technology. NR.

LOS ANGELES

In 1913 the city of Los Angeles completed a 200-mile-long aqueduct to carry water from the Owens Valley to the growing municipality. After World War I the city began building two large dams in the Los Angeles region to store water from the aqueduct. Both dams were designed by William Mulholland (1855–1935), chief engineer for the Los Angeles Bureau of Water and Power, and both were concrete curved gravity dams. The Mulholland Dam was built in the hills above Hollywood and provided a distribution reservoir for the rapidly developing city. Two years later the similarly designed St. Francis Dam in the St. Francisquito Canyon, about 40 miles north of Los Angeles, began impounding water, and for a short time both of Mulholland's designs were in operation. But on March 12, 1928, the St. Francis Dam collapsed under the load of a full reservoir, killing more than 400 people in the Santa Clara River Valley. The disaster, one of the worst in California history, brought Mulholland's engineering

■ **Mulholland Dam**
In the Hollywood Hills
East of U.S. Route 101, near
Lakeridge Drive
William Mulholland
1924

Mulholland Dam after earthfill was added on the downstream side, c. 1940.

career to an abrupt halt. It also raised fears about the safety of the previously built Mulholland Dam, although the St. Francis Dam failed largely because of poor foundation conditions. Geological studies verified the strength of the Mulholland Dam's foundations, but in the early 1930s the city bowed to public pressure and covered the downstream side of the dam with a large amount of earthfill. This gesture served no structural purpose but made the dam look stronger (or at least less tall). The earth-covered, 208-foot-high Mulholland Dam is still in use as part of the city's water supply system.

MATHER

■ **O'Shaughnessy (Hetch Hetchy) Dam**
Across the Toulumne River
In Yosemite National Park,
north of State Route 120
M. M. O'Shaughnessy
1923, 1938

By the early 20th century San Francisco had largely exhausted most nearby supplies of water and had begun exploring sources hundreds of miles away in the Sierra Nevada. The idea was to tap into a mountain stream and deliver water to the city through a lengthy tunnel-and-aqueduct system. By 1913 the city and its consultants, including the prominent hydraulic engineer John R. Freeman, were convinced that the best and most economical source of water was the Toulumne River in the central part of the state. The only problem was that the ideal dam site would inundate the Hetch Hetchy Valley in the northern part of Yosemite National Park. The proposed dam prompted one of the first environmental battles in the United States, but the initial controversy ended when the federal government agreed with the city and approved the dam as an economic necessity. Construction of the curved gravity dam proceeded under the direction of City Engineer M. M. O'Shaughnessy and became operational in 1923. In 1938 it was raised to its present height of 430 feet. The dam now carries the name of its original designer and remains a major source of water for San Francisco.

In a surprise development in mid-1987, Secretary of the Interior Donald Hodel made a startling proposal that the dam be torn down and the valley reclaimed. The tradeoff, in order to continue supplying San Francisco's water needs, would be to resume construction on the long delayed Auburn Dam on the American River east of Sacramento; this project was abandoned in the 1970s

Hetch Hetchy Dam, c. 1930, before this curved gravity structure was raised to its ultimate height.

when environmental and earthquake-stability questions were raised. Secretary Hodel's suggestion provoked some mild interest but much skepticism from conservationists and local politicians alike. Certainly the removal of the dam would make history, as few public works projects of this magnitude are ever reversed. But if the dam's abandonment does come to pass, it will take scores, if not hundreds, of years for the natural ecology of the Hetch Hetchy Valley to reestablish itself.

OAKLAND

Frequently overshadowed by its more famous neighbor spanning the Golden Gate (1937), the San Francisco–Oakland Bay Bridge is an impressive structure in its own right. Built by the California Toll Bridge Authority in conjunction with the California State Department of Public Works and the Reconstruction Finance Corporation, the bridge has an overall length of more than eight

■ **San Francisco–
Oakland Bay Bridge**
Across San Francisco Bay
On Interstate 80
C. H. Purcell and
Glenn B. Woodruff
1936

San Francisco–Oakland Bay Bridge. Above and left: Bridge under construction, c. 1935. Below: Inside the hollow midspan anchorage.

San Francisco–Oakland Bay Bridge. Left: Suspension span between San Francisco and Yerba Buena Island. Right: Cantilever truss leading from the island to Oakland.

miles and two main components. The East Bay crossing connects Oakland with Yerba Buena (Treasure) Island in the middle of San Francisco Bay. This portion consists of a 1,400-foot-long main cantilever span, five through trusses and 14 deck trusses; the total length, not including approach spans on the Oakland shore, is more than 11,000 feet. The West Bay crossing connects the island with San Francisco and consists primarily of two suspension spans with a total length of more than 9,200 feet. Between the suspension spans there is a huge concrete anchorage structure that connects the two together. The remainder of the bridge consists of lengthy approach spans in Oakland and San Francisco and a 1,800-foot-long tunnel and fill section on Yerba Buena Island. The structure is gradually becoming another national symbol of the Bay Area and recently beat the Golden Gate Bridge in celebrating its 50th anniversary by a half year. ASCE.

PASADENA

■ **Colorado Street Bridge**
Across the Arroyo Seco
On Colorado Boulevard
Waddell and Harrington
1913

Cutting through the heart of one of southern California's most developed areas, the Arroyo Seco is a deep, natural canyon separating Pasadena from Los Angeles. In the late 19th century a wooden truss bridge provided access across the normally dry arroyo. By 1912, however, Edwin Sover, director of Pasadena's board of trade, recognized that this relatively low-lying structure was not well suited for carrying the ever-increasing automobile traffic between the two cities. He began actively promoting a high-level, reinforced-concrete bridge and within a short time had received commitments from both cities to finance about $200,000 worth of construction. Sover commissioned the nationally known engineering firm of Waddell and Harrington to design a multispan reinforced-concrete arch bridge for the site. Although J. A. L. Waddell (1854–1938) had never worked with concrete before (his specialty was steel trusses), the firm developed an impressive design that ultimately served as the basis for the bridge. Waddell's design, however, required more money to build than Sover could extract from the two cities, and, over Waddell's objections, it was altered to reduce construction costs. This alteration, which involved building the bridge's eastern end on a pronounced

curve to place the structure on a bedrock ridge, reduced the amount of concrete required in the piers and arches and made it financially feasible for construction to proceed. Rehabilitation of the 1,467-foot-long, nine-span arch bridge is now being planned and will probably involve replacing the traffic deck. However, the original arches should remain in place and the bridge will continue to serve the region's transportation needs. NR.

Open-spandrel design of the Colorado Street Bridge.

RANCHO SANTA FE

In the early 20th century the Atchison, Topeka and Santa Fe Railway purchased the Rancho Santa Fe (a ranch first settled under Mexican rule), about 20 miles north of San Diego, to raise eucalyptus trees for railroad ties. The eucalyptus project proved a bust, but in 1916 the company joined with a local real estate entrepreneur, Col. Ed Fletcher, to develop an irrigation project on the land. The railroad financed construction of the Lake Hodges Dam, a 136-foot-high, multiple-arch dam on the San Dieguito River, and played a key role in the operation of the San Dieguito Irrigation District. In 1925 the city of San Diego purchased the Lake Hodges Dam and the rights to much of the water stored behind it. The city supposedly strengthened the dam against earthquakes in the mid-1930s by adding concrete stiffeners between the buttresses. The dam remains a vital component of the city's water supply system and a monument to a railway company's efforts to develop the West's scarce water resources.

■ **Lake Hodges Dam**
Across the San Dieguito River
Adjacent to County Route 6 (Del Dios Highway)
John S. Eastwood
1918

Lake Hodges Dam withstanding a flood before its buttresses were "strengthened" in 1936.

Delta Queen passing under the Tower Bridge, c. 1936.

SACRAMENTO

■ **Tower Bridge**
Across the Sacramento River
On Capital Mall Avenue
(State Route 275)
Bridge Department,
California Division of
Highways
1935

With the discovery of gold in northern California in the late 1840s, Sacramento quickly developed into a major commercial center. Ocean-going ships traveled up the Sacramento River and established a strong tradition of inland navigation in the region. Use of the river as a transportation route subsequently required construction of numerous movable bridges to provide clearance for ships. Among the most visually impressive of these movable spans is the Tower Bridge in the heart of Sacramento. This structure is distinguished by a 200-foot-long, vertical-lift span supported on two large Moderne-style towers. The architectural design was provided by Alfred Eichler of the Office of the State Architect. The Tower Bridge replaced an earlier swing span on the site that carried both railroad and highway traffic. For almost three decades the new bridge also carried a rail line, but it was removed in 1963 when the Sacramento and Northern Railroad obtained track rights on the nearby Southern Pacific Railroad bridge. Otherwise, except for a change in color from silver to ochre, the Tower Bridge is substantially unchanged and remains one of the most visually prominent structures in California's capital. NR.

SAN DIEGO

■ **Sweetwater Dam**
Across the Sweetwater River
½ mile north of
State Route 17
James D. Schuyler
1888

In the early 19th century Spanish settlers established a mission at San Diego, but little development of the community occurred until the end of the century. The prime deterrent to increased settlement was a lack of water, a problem that real estate developers were forced to confront if they wished to entice residents to the region. In the mid-1880s the San Diego Land and Town Company started building the Sweetwater Dam to provide water for irrigation and domestic use on land south of the city. The Sweetwater River is a torrential stream that carries a large amount of water in the spring but is essentially dry for the rest of the year, so the company sought to build a large dam capable of impounding all the spring runoff for use during the lengthy dry season. After an aborted attempt to work with F. E. Brown, engineer for the 1884

Sweetwater Dam being overtopped in 1895. The arch design survived with only minor damage.

Bear Valley Dam near San Bernardino, the company hired James D. Schuyler (1848–1912) to design an arch dam, 98 feet high with a maximum thickness of 46 feet. With these cross-sectional dimensions, the structure was too thin to function as a gravity dam. In 1895 heavy floods overtopped the Sweetwater Dam and provided dramatic evidence of the great strength of arch structures under hydrostatic load. Since that time, the spillway capacity of the dam has been increased to prevent further overtopping. The dam continues in active use as a part of the city's municipal water supply system.

SAN FRANCISCO

The Alvord Lake Bridge is the oldest reinforced-concrete bridge in the United States. Built with an extremely modest 20-foot span, it has a width of 64 feet. The reinforcing system consists of twisted iron rods imbedded into the concrete, a type of reinforcement that was a standard feature of Ernest Ransome's extensive work in reinforced-concrete construction. Because of its location within Golden Gate Park, the bridge was given special architectural features including an imitation stone surface and numerous concrete "stalactites" that hang from the interior curve of the arch. Although not built on a

■ **Alvord Lake Bridge**
In Golden Gate Park
Ernest Ransome
1889

grand scale, the reinforced-concrete technology used for the Alvord Lake Bridge served as a significant precursor for later work in the field. Ransome left California in the mid-1890s and became a major proponent of reinforced-concrete building construction in the East. He is well known to architectural and technological historians for his early reinforced-concrete patents and structures such as the Leland Stanford, Jr., Museum (1889) at Stanford University, the Pacific Coast Borax Company plant (1898) in Bayonne, N. J., and the United Shoe Machinery Company factory (1905) in Beverly, Mass. ASCE.

■ **Golden Gate Bridge**
Across the Golden Gate
at the mouth of San
Francisco Bay
On U.S. Route 101
Joseph B. Strauss, with
Charles Ellis and Leon
Moisseiff
1937

Golden Gate Bridge during
construction, c. 1935,
showing the cables without
the deck truss.

When Sir Francis Drake sailed up the Pacific coast in the late 16th century, coastal fog apparently kept him from discovering the entrance to San Francisco Bay. But ever since then few visitors to San Francisco have missed out on seeing the Golden Gate, which connects the bay with the Pacific Ocean. For many years people dreamed of joining San Francisco and the counties to the north with a permanent structure across the Golden Gate, but treacherous currents and the necessity for deep foundations made the idea of building piers between the two shorelines seem impossible. To cover the distance a bridge needed a clear span of at least 4,000 feet. The difficulties in funding a structure of this size precluded serious consideration of the project until after World War I.

In 1918 the city of San Francisco began to investigate potential sites for a bridge and, as part of this process, invited Joseph B. Strauss (1870–1938) to submit a feasibility report and a preliminary design. Strauss was a relatively prominent engineer known for his work with movable bridges. Even more important, Strauss was a close professional colleague of San Francisco's city engineer, M. M. O'Shaughnessy. Strauss's initial proposal, calling for a rather bizarre cantilever-suspension structure, was not accepted but impetus for the project was beginning to grow. In 1923 the state legislature authorized the Golden Gate Bridge and Highway District to sell bonds to finance construction, and more serious design investigations were undertaken. An important impediment to the project, however, was the considerable amount of public skepticism about the bridge. In its May 2, 1925, issue *The Wasp*, a local publication, summed up some of the thoughts on people's minds:

Discriminating persons believe that such a structure would prove an eye-sore to those now living and a betrayal of future generations, for whom the present generation is a trustee. A bridge of the size projected — the plans call for towers 800 feet high — would certainly mar if not utterly destroy the natural charm of the harbor famed throughout the world.

In retrospect, it may seem incredible that people were opposed to building the bridge, but in the context of the times their objections were not so unreasonable. After all, it represented a huge public expenditure that many people believed could not be repaid easily through toll revenues.

Golden Gate Bridge. Left: Looking toward the San Francisco peninsula from the top of the tower. Below: Section of cable similar to that used for the Golden Gate Bridge. Bottom: Bridge from the Marin County shoreline.

Worker wrapping a cable of the Golden Gate Bridge.

Finally, after years of planning and political maneuvering, Strauss was formally named chief engineer for the project in 1929. However, construction did not begin for another three and a half years. Work on the 4,200-foot-long clear span suspension bridge took four years, and it was finally opened for traffic in May 1937. Data on the structure's dimensions and superlative statistics cannot begin to convey the affection that Americans from all parts of the United States have for the bridge. Regardless, it is worth noting that the towers are 746 feet tall; each of the two cables has a diameter of 36⅜ inches (and they are designed to carry a total load of 160 million pounds); the clearance above the water is 220 feet at midspan; and the total overall length of the structure, including approach spans, is more than 8,900 feet. The architectural treatment for the towers was designed by Irving Morrow under Strauss's direction.

In a 1987 celebration commemorating the 50th anniversary of the span, more than a quarter of a million people turned out to walk across the structure. Although officials had anticipated a large turnout, the size of the crowd exceeded all expectations and prompted concern that the excessive loads might overstress the cables. However, the Golden Gate Bridge survived the celebration, and it remains an icon that occupies a special place in America's cultural consciousness. ASCE.

SAN MATEO

■ **Crystal Springs (San Mateo) Dam**
Across San Mateo Creek
On Crystal Springs Road, ½ mile west of Interstate 280
Hermann Schussler
1888

Before the early 20th century the city of San Francisco received its water from the Spring Valley Water Company, a private firm operating under a city franchise. In the 1880s this company undertook a major expansion program and, under the supervision of Hermann Schussler (1842–1919), began building a concrete curved gravity dam on the peninsula south of San Francisco. Born and trained in Switzerland, Schussler was one of the most prominent hydraulic engineers in 19th-century California, and he was associated with many early gold-mining operations. The Crystal Springs Dam, 146 feet high with a maximum base width of 176 feet, was described by dam engineer and author James D. Schuyler

Crystal Springs Dam in 1896, when members of the American Society of Civil Engineers visited the massive concrete gravity structure.

in 1909 as "doubtless the most enormous mass of masonry of any work in the West, if not in the entire United States." Schuyler also called it "among the highest and most costly dams in the world." Aggregate for the concrete came from local quarries, while the sand was shipped in from San Francisco's North Beach. The dam was poured in place in a series of interlocking blocks containing about 250 cubic yards of concrete. The structure became the object of engineering acclaim after it successfully withstood the famous San Francisco earthquake of April 1906. In 1911 the company added eight feet to its height, but, otherwise, the dam remains practically unchanged from its original design. It is now part of San Francisco's municipally owned water supply system.

SUMMIT CITY

The Shasta Dam, a 602-foot-high, 3,400-foot-long, concrete curved gravity structure, was undertaken as part of the Central Valley Project, the first of the Bureau of Reclamation's large-scale activities in California. This project involved building a major dam on the upper Sacramento River to store floodwaters that would otherwise flow into San Francisco Bay. Water released from behind the Shasta Dam flows in the river channel until reaching the delta below Sacramento; from there it is pumped south for use in irrigating land in the lower San Joaquin Valley. Its partner, the Friant Dam (1942), in turn, diverts water from the upper San Joaquin River south for use in irrigating the fertile yet dry regions south of Fresno near Bakersfield. In essence, the Central Valley Project takes excess water in the northern part of the state and allows it to be used for irrigation in the area around the lower San Joaquin River. This facilitates the transfer of San Joaquin River water to arid land in the southern part of the state. By "swapping" water on a huge geographical scale, it became possible to open up large tracts of otherwise arid land for agriculture. When completed, the Shasta Dam was the third most massive concrete dam in the West after the Hoover and the Grand Coulee dams. It remains in active use.

■ **Shasta Dam**
Across the Sacramento River 10 miles east of Interstate 5
U.S. Bureau of Reclamation
1945

Shasta Dam, with its overflow spillway in the middle of the structure.

VALLEJO

■ **Carquinez Strait Bridge**
Across the Carquinez Strait
On Interstate 80
David B. Steinman and
William H. Burr
1927

Above: Original Carquinez
Strait Bridge, c. 1935, before
construction of the parallel
span.

The huge Central Valley of California is drained by the Sacramento River to the north and the San Joaquin River to the south. These two rivers join together in the delta region about 50 miles southwest of Sacramento and then flow westward toward San Francisco Bay. Just before entering the bay, the waterway formed by the two rivers passes through a relatively narrow strait. Known as the Carquinez Strait, this location became the site of a major highway bridge built in the 1920s. Constructed by the privately financed American Toll Bridge Company, the bridge consisted of a 3,350-foot-long steel cantilever through truss with two main spans of 1,100 feet each. Construction involved building a center pier on bedrock that lay 50 feet below the water surface.

The bridge proved to be a major success, and in 1958 the California State Highway Department constructed a new cantilever structure of dimensions similar to the original 1927 bridge. Built 200 feet apart, the two structures now work in tandem to carry highway traffic on Interstate 80; the 1927 bridge handles all southbound cars and trucks while the 1958 bridge handles northbound traffic.

■ ■ ■ ■ ■ ■ ■ COLORADO ■ ■ ■ ■ ■ ■ ■

ANTONITO

■ **Costilla Crossing Bridge**
Across the Rio Grande
14 miles east of U.S.
Route 285
Wrought Iron Bridge
Company
1892

In 1891 the Colorado state engineer solicited bids for the design and construction of a permanent bridge across the Rio Grande near the Colorado–New Mexico border. More than 38 proposals were submitted, and, after due deliberation, the Wrought Iron Bridge Company received a contract to build a two-span structure at the site in Conejos County for $8,400. The design selected for construction was not the usual Pratt configuration; instead, it was a relatively rare example of the Thacher truss patented in 1884 by Edwin Thacher (1840–1920). Thacher served as bridge engineer for Andrew Carnegie's Keystone Bridge Company in the 1870s and later became famous as an early American pioneer in reinforced-concrete bridge construction. But in the 1880s he devoted considerable attention to devising a truss design that would, in his mind, reduce the effect of temperature stresses on the various members.

The resulting Thacher truss represents a kind of hybrid between the double-intersection Pratt and Warren designs. It is distinguished visually by diagonal compression members located at the center of each span and diagonal tension members that extend from the connection between the top chord and the inclined end post. The Wrought Iron Bridge Company built a small number of Thacher trusses in the 1880s and 1890s, but the design never became very popular, probably because it required rather complex cast-iron connector blocks for the diagonal compression members. Two other Thacher trusses — in Michigan and Virginia — survived into the 1980s, but the Costilla Crossing Bridge is the only two-span Thacher truss known to exist in the United States. Located in an extremely remote part of the state, it is among the oldest unaltered bridges in Colorado still carrying highway traffic. NR.

Two-span Costilla Crossing Bridge across the Rio Grande on a wintry day.

ASPEN

The town of Aspen boomed in the 1880s, and both the Colorado Midland and the Denver and Rio Grande railroads extended their lines to serve the burgeoning mining community. Among the longest crossings built for these branch lines was the trestle at Maroon Creek. By early 1888 the Colorado Midland Railroad completed the 651-foot-long bridge, which consists of 20 spans, each with a maximum length of 30 feet. The iron trestle also includes nine riveted towers supported on stone footings, the tallest tower being more than 90 feet high from foundation to the deck. In 1919 the railroad abandoned its Aspen line, and in 1929 the state highway department formally adapted the Maroon Creek Trestle for highway use. Aside from alterations to the road deck, the bridge retains most of its 19th-century design and survives as the oldest and largest railroad trestle in the state now used for automobile and truck traffic. In a recent inventory of historic bridges undertaken for the Colorado Department of Highways, western bridge historian Clayton Fraser lauded it as "one of the last remaining iron/steel multiple span trestles erected in the 19th century for Colorado's narrow gauge mountain railroads." NR.

■ Maroon Creek Trestle
Across Maroon Creek
On State Route 82, 1¼ miles west of Aspen
George S. Morison
1888

CANON CITY

■ Royal Gorge Bridge
Across the Arkansas River
At Royal Gorge, 5 miles
northwest of Canon City
George F. Cole
1929

Most bridges in the United States were built as part of a larger highway system, a railroad line or an aqueduct. But a few, such as the Royal Gorge suspension bridge, were intended to be, first and foremost, tourist attractions. In 1906 Congress transferred control of the Royal Gorge (a deep valley in the upper reaches of the Arkansas River watershed) in southern Colorado to Canon City so long as the municipality used the property for recreational purposes. In the late 1920s the city chose to heighten public interest by authorizing construction of a steel suspension bridge that would allow "aerial views" of the visually spectacular gorge. Between June and December 1929 the Royal Gorge Bridge and Amusement Company constructed the 1,220-foot-long bridge with a clear span between piers of 880 feet. Rising more than 1,050 feet above the Arkansas River (it is touted as the highest bridge in America), the bridge consists completely of structural material fabricated in Colorado. Aside from periodic refurbishment of the wooden deck, the bridge remains essentially unaltered. Tourists visiting Royal Gorge can ride on an inclined railway from the bridge site to and from the bottom of the canyon. NR.

DECKERS

■ Cheesman Dam
Across the South Platte
River
Near State Route 67
Charles Harrison and Silas
Woodward
1904

Cheesman Dam nestled in
the Rocky Mountains in
1904.

In 1898 the Denver Union Water Company, acting through its subsidiary, the South Platte Canal and Reservoir Company, began construction of a rockfill storage dam on the upper reaches of the South Platte River. Its purpose was to increase the water supply for Denver, Colorado's largest city. In May 1900 a spring flood overtopped the partially completed dam and washed away practically the entire structure. The company quickly reassessed the situation, and by September it had started work on a masonry curved gravity dam at the same site. Completed four years later, the Cheesman Dam is 232 feet high and has a maximum base width of 176 feet. Granite used for the imposing structure was quarried 2,000 feet from the dam site. Located about 40 miles southwest of Denver, the dam continues to provide water for the burgeoning metropolitan area. ASCE.

FORT MORGAN

Among the most visually attractive types of reinforced-concrete bridges built in the United States is the rainbow arch bridge, popularized by the Marsh Engineering Company of Des Moines, Iowa. Patented by James Marsh (1856–1936) in 1912, although others had previously built structures almost identical in design, the rainbow arch design consists of open-spandrel arches that rise above the traffic deck. In essence, it is similar to a bowstring arch truss except that all the major structural components consist of reinforced concrete. Built by the Colorado highway department and Morgan County in 1922–23, the Fort Morgan Bridge is among the largest rainbow arch bridges in the United States and the only example of its type in Colorado. It consists of 11 arches, each with a 90-foot span, and has a combined length of more than 1,100 feet. In 1963 a second bridge was built nearby to help relieve traffic on the 1923 structure. Aside from some remedial work on the three southernmost arches, the historic bridge still retains its structural integrity and remains an important component in the regional transportation system.

■ **Fort Morgan Bridge**
Across the South Platte River
On State Route 52, 1 mile north of Fort Morgan
Colorado State Highway Department
1923

Left: Rainbow arches of the Fort Morgan Bridge. Right: Fruita Bridge, with its pin-connected Parker trusses.

FRUITA

In 1906 the state of Colorado, the county of Mesa and the town of Fruita agreed to share the cost of building a steel truss bridge across the Colorado River about 20 miles east of the Utah state border. Completed in 1907, the Fruita Bridge is a three-span, pin-connected Parker through truss with an overall length of 472 feet including approaches. Each of the main truss spans is 155 feet long. The highway bridge remained in service for almost 65 years before being bypassed by a new structure in 1970. Because the new bridge was built along a different alignment, its construction did not require the demolition of the 1907 structure. Today, the historic Fruita Bridge is accessible only to pedestrians and it carries no vehicular traffic. It is considered among the oldest and most attractive truss bridges in the state and was featured on the cover of the Historic Bridges of Colorado inventory recently published by the state department of highways.

■ **Fruita Bridge**
Across the Colorado River
Adjacent to State Route 340, on County Route 17.50
M. J. Patterson Construction Company
1907

■ ■ ■ ■ ■ ■ ■   GUAM   ■ ▪ ■ ■ ■ ■ ■

AGANA

■ Agana Spanish Bridge
Across the old Agana River
diversion channel
At Aspinwall Street and
Route 1
c. 1800

When Ferdinand Magellan (or, more properly, his crew) completed the first voyage around the world in 1518–21, one of his landfalls in the mid-Pacific Ocean was the island of Guam. This stopover marked the beginning of a Spanish presence on the island that lasted for hundreds of years until the 1898 Spanish-American War. Evidence of Guam's Spanish colonial heritage can be found in a single-span stone arch bridge located near the center of Agana, the island capital. Toward the end of the 18th century, the Spanish government diverted part of the Agana River through the center of the city and, to provide access over the diversion canal, built a small highway bridge as part of this civic development project. During the American invasion of the island in World War II (it was held by the Japanese from December 1941 until July 1944), the bridge experienced some damage but not enough to destroy its historical integrity. Since then the project to divert river water through the city has been abandoned, and the bridge no longer serves any transportation function. It is preserved today as one of the island's oldest structures. NR.

AGAT

■ Taleyfac Spanish Bridge
Across the Taleyfac River
Adjacent to Route 2
c. 1870

Taleyfac Bridge on the
southwest coast of Guam.

In the late 18th century the Spanish government constructed a road down the west coast of the island connecting the capital of Agana with settlements in the southern part of the island. Bridges on the road were originally built of wood, but in the mid-19th century Gov. Felipe de la Calderan began a program of replacing deteriorated wooden spans with more permanent stone structures. Among these new bridges was the 36-foot-long, two-span, masonry arch bridge over the Taleyfac River. This structure is no longer used to carry highway traffic, but it remains in place adjacent to new Route 2. NR.

■ ■ ■ ■ ■ ■ ■   HAWAII   ■ ■ ■ ■ ■ ■ ■

HANALEI, KAUAI

The north coast of Kauai is among the most beautiful areas in the United States. Located only a few miles from Kalulau Lookout (the wettest place on Earth), the community of Hanalei is a tropical paradise at the center of life on the north coast. Aside from ships entering Hanalei Bay, the only way to reach Hanalei is via State Route 56, a simple two-lane road extending north of Kapaa. The 110-foot-long Pratt through truss bridge that carries the road over the Hanalei River became the focus of controversy several years ago. Completed in 1912 and supplemented by a Warren pony truss in the 1960s, the span suffered serious structural deterioration over many ·years. The state transportation department decided to replace it, but the community opposed construction of a new bridge and sought a way to rehabilitate the older structure. After years of discussion an agreement was recently reached to retain the original 1912 structure in place but to have the supplemental Warren truss carry all of the traffic loads. This plan, developed by bridge renovation specialist Abba Lichtenstein and his engineering firm, allows for both a safe highway system and the retention of a structure that melds in easily with the historic and cultural landscape of Hanalei. If you've been looking for a reason to visit the Garden Isle of Kauai, it would be difficult to pick a more lovely spot to see firsthand a successful example of historic bridge rehabilitation.

■ Hanalei River Bridge
Across the Hanalei River
On State Route 56
Hamilton and Chambers
1912, c. 1965

Above: Hanalei Valley, with the bridge in the lower foreground. Left: Hanalei Bridge showing the original Pratt through truss and the later Warren pony truss addition.

KAPAA, KAUAI

During much of the 19th century the Hawaiian Islands existed as a constitutional monarchy with economic ties to both Great Britain and the United States. An 1876 treaty granted the United States military privileges in the islands in return for allowing sugar to be exported to American markets duty free. Despite this, Britain still maintained an economic presence so strong that when the Hawaiian legislature authorized construction of a bridge across the mouth of the Wailua River on Kauai's eastern shore, it placed the order with Alexander Findlay and Company of Motherwell, near Glasgow, Scotland. After delivery of the three-span truss in the early 1890s, political problems caused by the overthrow of the monarchy and the establishment of the Hawaiian Re-

■ Opaekaa Truss Bridge
Across Opaekaa Stream
On Opaekaa Road
Alexander Findlay and Company
c. 1895, 1919

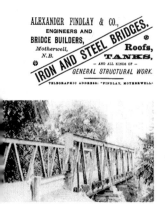

Opaekaa Bridge and an advertisement from 1888 for the Scottish company that fabricated it.

public in 1893 delayed erection of the bridge. But through the efforts of George Wilcox, owner of Kauai's large Grove Farm plantation, the highway bridge was in use by 1895. It remained in operation until 1919, when the county built a new reinforced-concrete bridge over the Wailua River. At that time one span of the Scottish truss found a second career a few miles inland across the Opaekaa Stream.

The Opaekaa Bridge is a single-span, riveted Warren pony truss. It is 73 feet long and is supported on lava-rock masonry abutments. The boxy dimensions of the truss and the form of its riveted connections are unlike those found in any late 19th-century American bridge. But lest anyone doubt the age and origin of the Opaekaa Bridge, it still carries a cast-iron nameplate listing the builders as "Alex. Findlay & Co. Bridge Builders, Motherwell near Glasgow" and the fabrication date "1890." It is the only known British truss bridge in the United States. It also provides compelling physical evidence of Great Britain's economic influence in the Pacific region during the 19th century. NR.

■ ■ ■ ■ ■ ■ ■   IDAHO   ■ ■ ■ ■ ■ ■ ■

BOISE

■ **Arrowrock Dam**
Across the Boise River
5 miles east of State
Route 21
U.S. Reclamation Service
1916, 1937

Soon after the establishment of the Reclamation Service in 1902, agency engineers surveyed the Boise River Valley for reservoir sites. By 1911 intensive drilling to determine foundation conditions at the Arrowrock site was under way, and the following year work began on building a concrete curved gravity storage dam. For more than a decade the 349-foot-high structure was the tallest dam in the world until superseded by the Owhyee Dam in eastern Oregon. Water from the reservoir is used for irrigation projects in the Boise and Snake river basins and makes a vital contribution to the economy of southwestern Idaho. In 1936–37 the dam was raised five feet in height, and, because of spalling on the downstream side of the structure, much of it was covered with a new 18-inch-thick slab of reinforced concrete. Otherwise, the Arrowrock Dam has not changed much over the past 70 years, and it is still used for its original purpose. NR.

Arrowrock Dam. Left: View looking downstream during early construction.
Right: Aerial view looking upstream.

Upstream face of the Fish Creek Dam as it appeared in 1920.

CAREY

Mormon settlement in the West is usually associated with Salt Lake City and the state of Utah. Historically, however, the influence of the Church of Jesus Christ of Latter-day Saints has extended well beyond the borders of Utah. Church members, for example, pioneered in the settlement of southern Idaho, and this area still remains a strong bastion of the Mormon faith. Economic development in the region is closely tied to irrigation, and the Fish Creek Dam was built by Mormon farming interests shortly after World War I. The dam was designed by John S. Eastwood, a prolific dam engineer and non-Mormon from California. Impetus for the dam's construction came after Eastwood received the design commission for Salt Lake City's Mountain Dell Dam (1917). The Fish Creek Dam is a reinforced-concrete, multiple-arch structure consisting of 68 arches. It has a maximum height of 92 feet and is 1,725 feet long. The concrete in the structure is somewhat deteriorated, but the dam still contributes to irrigation in the Carey region. NR.

■ **Fish Creek Dam**
Across Fish Creek
8 miles north of U.S.
Route 93
John S. Eastwood
1920

MINIDOKA

The Minidoka Dam is the key structure in the Reclamation Service's Minidoka Project, which provides water and hydroelectric power to several irrigation districts along the Snake River in southern Idaho. The dam is an 86-foot-high earthfill and rockfill dam with a total length of 4,475 feet. Because of the Snake River's great flow, the structure incorporates a 2,400-foot-long concrete gravity spillway into its design. Along with four other storage dams in the Minidoka Project, the Minidoka Dam helps

■ **Minidoka Dam**
Across the Snake River
10 miles south of State
Route 24
U.S. Reclamation Service
1906

Minidoka Dam, with water flowing over the spillway.

Interior of the hydroelectric powerhouse at the Minidoka Dam.

irrigate more than one million acres of land. Its 1913 power plant is considered an important precedent for large federal power-generating facilities in the Pacific Northwest, including the Bonneville and Grand Coulee dams. NR.

MURTAUGH

■ **Milner Dam**
Across the Snake River
North of U.S. Route 30
A. J. Wiley
1904

In 1894 Congress passed a law allowing federal land to be conveyed to the western states in order to foster irrigation projects. Known as the Carey Act in recognition of its chief proponent, Sen. Joseph Carey of Wyoming, the significance of the law has been largely overshadowed by the 1902 Reclamation Act that founded the U.S. Reclamation Service. The Carey Act nonetheless significantly affected agricultural development in many western states. As noted by western historian Merle Wells, the most important of all Carey Act projects is the Milner

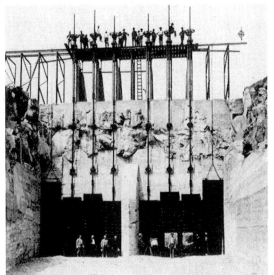

Above: Rockfill Milner Dam, with water passing over the spillway. Right: Diversion gates used to regulate the flow of water into the irrigation canals.

Dam and its extensive 160-mile canal system, located in southern Idaho. Constructed using private capital, the Milner Dam is an 86-foot-high rockfill dam that stretches for 2,320 feet across the Snake River. Since the completion of the dam and the associated Twin Falls Canal, more than 360,000 acres of land have become agriculturally productive. The dam remains in use today as a monument to the substantial accomplishments made possible by the largely unheralded Carey Act. NR.

■ ■ ■ ■ ■ ■ ■   MONTANA   ■ ■ ■ ■ ■ ■ ■

AUGUSTA

Located on a relatively isolated stretch of highway in the hills west of Great Falls, the Dearborn River High Bridge is a rare example of a half-deck Pratt truss. In basic terms, its design is not dissimilar to other pin-connected Pratt trusses, but this 160-foot-long structure is distinguished by having the deck located between the top and bottom chords. As a result, the top chord acts essentially as a type of guardrail parallel to the traffic deck. It is not known why such an unusual design was selected for this site or why a Cleveland-based company received the design commission. It still carries regional highway traffic.

■ **Dearborn River High Bridge**
Across the Dearborn River
On County Road 434, near Bean Lake
King Iron Bridge and Manufacturing Company
1897

Dearborn Bridge, showing its distinctive half-through truss design.

The Gibson Dam impounds water for the Sun River Irrigation Project in west-central Montana. It is a 190-foot-high, concrete arch dam that differs from the gravity designs often built by the Bureau of Reclamation during this period. The arch design, which requires less material, was probably selected because of the site's good bedrock foundations and its location in an extremely remote part of the state. The dam lay 23 miles from the nearest railroad, and construction required building a 28-mile-long transmission line to deliver electric power to the site. Although not located near any heavily traveled tourist routes, the Gibson Dam is accessible to travelers wishing to journey into the hinterlands west of Augusta.

■ **Gibson Dam**
Across the North Fork of the Sun River
20 miles northwest of U.S. Route 287
U.S. Bureau of Reclamation
1929

FORT BENTON

■ **Fort Benton Bridge**
Across the Missouri River
Adjacent to State Route 230
Milwaukee Bridge and Iron
Works
1888, 1925

The Fort Benton Bridge is the first all-iron vehicular truss bridge in Montana and the oldest surviving bridge of any type in the state. When the Great Northern Railroad reached Fort Benton in 1887, a group of local business-people formed the Benton Bridge Company to help facilitate trade with settlers in the Judith River Basin to the south. Completed a year later, the Fort Benton Bridge consisted of one Pratt through truss with a 75-foot span, three Baltimore through trusses, each with a span of 175 feet, and a 225-foot Pratt truss swing span. All of the trusses were pin connected. Officially the Missouri River was classified as navigable at Fort Benton, but the swing span did not operate until 1908, when a steamboat first required clearance to pass by. That same year a flood destroyed the swing span, and, after the Missouri River was officially declassified as navigable at this point, a temporary wood and iron truss was built in place of the swing span. In 1925 this was replaced with a permanent, all-steel Parker through truss. In a recent inventory of historic Montana bridges, engineering historian Fred Quivik proclaimed the Fort Benton Bridge as "the most historically significant" span in the state because of its age and its role in fostering the economic development of Fort Benton. NR.

Fort Benton Bridge. Below:
Two of the Baltimore trusses.
Bottom left: Riveted iron
sheathing on a pier. Bottom
right: Bridge's deck, no
longer maintained for
highway traffic.

FORT PECK

Despite its name, the Missouri River is often considered the main stem of the Mississippi River, North America's largest waterway. Flowing from the Rocky Mountains, the Missouri's primary headwaters are in Montana and Wyoming. Fed by major tributaries such as the Platte River, it also drains large portions of North Dakota, South Dakota, Nebraska, Colorado, Kansas, Iowa and Missouri before joining the Mississippi River at St. Louis. Beginning with Indian traders and trappers in the early 19th century and continuing with the growth of Montana's mining fields, the Missouri served as a major conduit for inland navigation. By the early 20th century the upper Missouri River was no longer used for navigation, but the lower stretches between St. Louis and Kansas City still carried river-borne commerce.

In the late 1920s the Corps of Engineers received congressional authorization to study the Missouri and develop a comprehensive plan for navigation, flood control, irrigation and power production. In the early 1930s the Kansas City district of the corps recommended construction of a large earthen dam at Fort Peck in eastern Montana. This structure was to provide flood control for the lower Missouri and, by regulating flow in normally dry months, facilitate navigation up to Kansas City. With Franklin Roosevelt's election to the presidency in 1932, the Fort Peck Dam found support as a major public works project designed to boost America's ailing economy. With initial funding of more than $80 million from the federal Public Works Administration, construction of the 250-foot-high, 21,000-foot-long earthfill structure began in late 1933. During the next seven years, thousands of workers were brought to the dam site, and the project helped bolster the region's depressed economy. Having served as a precedent for several other earthen dams built downstream after World War II, the Fort Peck Dam is now the oldest corps-built dam on the Missouri River. The dam is also remembered because a Margaret Bourke-White photo of its concrete spillway graced the cover of the inaugural issue of *Life* magazine in 1936. NR.

■ **Fort Peck Dam**
Across the Missouri River
Adjacent to State Route 24
U.S. Army Corps of Engineers
1940

Above: Area surrounding the earthfill Fort Peck Dam.

MILES CITY

■ **Fort Keogh Bridge**
Across the Yellowstone River
On Local Road 56
W. S. Hewett and Company
1902

This impressive structure consists of two small approach spans and two 310-foot-long, pin-connected Pennsylvania through trusses. At the time of its construction, the nearest permanent crossing of the Yellowstone River was at Glendive, almost 80 miles downstream. The Fort Keogh Bridge was fabricated by W. S. Hewett and Company, a Minneapolis-based firm responsible for many other truss bridges built in Montana during the early 20th century. Following the conversion of the Fort Keogh Military Reservation into a livestock research station for the U.S. Department of Agriculture in 1924, the bridge also came under the control of the USDA. Practically unchanged, it is used to carry limited local highway traffic.

SNOWDEN

■ **Snowden Lift Bridge**
Across the Missouri River
On the Burlington Northern
Railroad right-of-way, near
State Route 200
Waddell and Harrington
1913

Located at the eastern edge of the state to allow for navigation along the upper Missouri River, the Snowden Bridge is the only vertical-lift bridge in Montana. Designed as a railroad structure for the Great Northern Railroad under the direction of J. A. L. Waddell and built by the American Bridge Company, it was among the longest movable bridges in the world at the time of its construction. Composed of three 275-foot-long, fixed-span riveted Parker through trusses and a 296-foot-long movable Parker truss, the Snowden Bridge was modified in 1926 to carry vehicular traffic. The structure is in good condition, although the lift span has not been used for many years and is now inoperable.

Hauser Lake Dam, a typical concrete gravity dam designed by the Boston firm of Stone and Webster.

YORK

In 1907 the Helena Power and Transmission Company completed a steel buttress dam at Hauser Lake designed to generate power for mining companies in nearby Helena. The structure was designed by J. F. Jackson, the same engineer responsible for the Redridge Steel Dam (1901) in Michigan. Unfortunately, the foundations at the Lake Hauser site were relatively porous, and the steel structure was not firmly keyed into bedrock. In April 1908 a large section of the steel dam washed away, causing the power company considerable economic anguish. In 1909 it began building a concrete gravity dam at the site and took great pains to ensure that it was supported on solid bedrock. The foundations were excavated by the Foundation Company of New York, while Stone and Webster handled the design of the 130-foot-high, 720-foot-long dam. It is now part of the Montana Power Company's generating system.

■ **Hauser Lake Dam**
Across the Missouri River
Adjacent to County
Route 280
Stone and Webster
1912

■ ■ ■ ■ ■ ■ NEVADA ■ ■ ■ ■ ■ ■

BOULDER CITY

Beginning in the early 20th century, engineers dreamed of tapping the Colorado River's hydroelectric power potential. At the same time, agricultural interests in the lower Colorado River Valley and southeastern California started seeking a means of reducing the scourge of floods and increasing the supply of water for irrigation. Although the Hoover Dam is often considered a major symbol of Roosevelt's New Deal, its origins lay in the efforts of large-scale farmers to develop California's Imperial Valley. After World War I, Arthur Powell Davis, then director of the U.S. Reclamation Service, began to promote construction of a large storage dam on the Colorado River. He was supported in this effort by a California congressman, Phil Swing, and senator, Hiram Johnson. But it was not until Los Angeles and other southern California cities developed plans to use the proposed dam as a source of water supply and hydroelectric power that support for the project gained momentum in Congress. The newly formed Southern California Metropolitan Water District joined forces with

■ **Hoover Dam**
Across the Colorado River
On U.S. Route 93
U.S. Bureau of Reclamation
1935

Hoover Dam, Boulder City, Nev. Top left: Black Canyon, site of the dam. Top right: Dam in its geological setting. Above: Workers using jackhammers to prepare the rock abutments. Above right: Spillway shortly after completion. Right: Dam with Lake Mead filled to the brim.

groups in the Imperial Valley, and finally, in December 1928, President Calvin Coolidge signed the Boulder Canyon Project Act, a name derived from the proposed location, authorizing construction to proceed.

Snugly fit into the hard rock canyon 30 miles southeast of Las Vegas, Nev., the Hoover Dam rises 726 feet above its foundations and is 660 feet thick at its deepest elevation. Extending more than 1,200 feet along the crest, the structure is built with a pronounced upstream curve. The dam appears to rely on arch action to resist the hydrostatic load imposed by Lake Mead. In reality, its dimensions are so ample that it could have been constructed straight across the canyon and still have maintained stability. Given that it was by far the highest dam in the world when built, however, the extreme conservatism of the design is understandable.

The Bureau of Reclamation took responsibility for the dam's design, but, because of its huge scale, construction was handled by a unique consortium of private firms known as Six Companies, Inc. Started in 1931, the project represented the largest contract ever awarded in the United States — more than $31 million. Technical features were determined by the bureau's engineering staff, but the Moderne style of the design was the work of Gordon Kaufmann, a Los Angeles architect who had never before designed a large-scale engineering project.

The original site of the dam proved unacceptable because of a fault discovered during foundation exploration; thus, the site was subsequently shifted about 10 miles downstream to a location in Black Canyon. Meanwhile, the dam had become known as the Hoover Dam, in honor of President Herbert Hoover. As the structure rose in Black Canyon, the Roosevelt administration changed the name to the less politically charged Boulder Dam. In the late 1940s the U.S. Congress

Left: Movable platform used to carry men and materials around the dam site. Right: Moderne-style intake towers directly upstream from the Hoover Dam.

changed the name back to Hoover.

Construction began with the raising of two temporary cofferdams that enabled "underwatering" of the dam site. This allowed the contractors to excavate down to bedrock under dry conditions. The flow of the Colorado River was then diverted through two large tunnels driven into the walls of the canyon. These tunnels were later incorporated into the spillway system. With dry conditions ensured by the cofferdam-tunnel system, construction proceeded at a rapid pace, especially when the Roosevelt administration began to promote the project as part of its work relief program. To support the work force, a complete town known as Boulder City was erected in the desert near the site. Concrete was poured on an around-the-clock basis, and, to remove the huge amounts of heat released by the concrete as it hardened, pipes carrying cold brine were placed throughout the structure. When construction of the dam proper ended in May 1935, it contained more than 3,250,000 cubic yards of concrete.

In 1937 the dam's hydroelectric plant began to generate power for transmission to southern California. Much of this power is used to pump water for the Southern California Metropolitan Water District's Colorado Aqueduct. Today, the plant's generating capacity is 1,376,250 kilowatts, and power is transmitted to California, Nevada and Arizona. Water stored behind the dam, which has a total capacity of 12 trillion gallons, is used for agricultural purposes in the Imperial Valley and industrial and domestic purposes in the Los Angeles area. On completion of the Central Arizona Project in the late 1980s, water stored behind the dam will also make its way to the greater Phoenix-Tucson region. NR, ASCE.

RENO

■ **Virginia Street Bridge**
Across the Truckee River
On Virginia Street
John B. Leonard
1905

Virginia Street Bridge,
c. 1910.

The Truckee River flows east from the Sierra Nevada in California and passes through a substantial part of west-central Nevada before entering Pyramid Lake northeast of Reno. The first structure at the Virginia Street site was a crude wooden toll bridge built in 1859 to handle the considerable traffic to the rich Comstock mines of Virginia City. It was this bridge that formed the locus for the original settlement of Reno. The original bridge and two later wooden bridges were washed out. In 1877 the city built an iron bowstring arch truss at the site, but by 1905 this bridge was considered inadequate and was

replaced with the present structure designed by John Leonard. Born in Michigan in the mid-1860s, Leonard came to San Francisco in 1889 to begin practice as a civil engineer. He later became one of the West's most prolific concrete-bridge builders, and the Virginia Street Bridge represents his first major commission. The Virginia Street Bridge is a two-span, reinforced-concrete arch bridge that extends 146 feet in length. Construction of the bridge began in July 1905 and was completed by mid-November. Although some surface spalling of the concrete has occurred, the structure maintains its historical integrity and still carries traffic in downtown Reno. NR.

SILVER SPRING

The Lahontan Dam is a 162-foot-high, 5,400-foot-long earthfill structure that impounds irrigation water for land in the Fallon area. It was the first large dam built as part of the Reclamation Service's Truckee-Carson Project, now the Newlands Project, which was inaugurated with the Derby Diversion Dam (1905). Water from the dam is also used to generate hydroelectric power for commercial and domestic users in the surrounding region. Visually, the design is distinguished by two overflow concrete spillways that cut through the earthfill embankment. NR.

■ **Lahontan Dam**
Across the Carson River
Adjacent to U.S. Route 50
U.S. Reclamation Service
1915

Above: Downstream side of the earthfill Lahontan Dam. Left: Water passing over one of the stepped concrete spillways.

■ **Derby (Truckee-Carson)**
Diversion Dam
Across the Truckee River
Adjacent to Interstate 80
U.S. Reclamation Service
1905

Above: Congressional party
gathering to watch the initial
opening of the Derby Dam
headgates in 1905.

U.S. Senator Francis G. Newlands of Nevada served as a leading legislative advocate for the 1902 Reclamation Act, which authorized establishment of the Reclamation Service; in fact, the bill is often referred to as the Newlands Act. Consequently, it should not be surprising that the initial project undertaken by the new federal agency was in the senator's home state. The Derby Diversion Dam was the first structure built by the Reclamation Service and, as such, received more attention than its modest size would normally warrant. The dam is a 31-foot-high, 170-foot-long, concrete gravity structure supplemented by a shallow earth embankment 1,160 feet long. Water from the Truckee River is diverted into a canal that carries it several miles into the lower Carson River watershed, where it is used for irrigation. Later supplemented by several other dams in the region, including the Lahontan Dam (1915), that are now collectively known as the Newlands Project, the Derby Dam is little changed and still provides support for the region's agricultural production. NR.

■ ■ ■ ■ ■ ■ OREGON ■ ■ ■ ■ ■ ■

BONNEVILLE

■ **Bonneville Dam**
Across the Columbia River
Adjacent to Interstate 84
U.S. Army Corps of
Engineers
1937

After the Mississippi, the Columbia River is the largest river in the continental United States (its annual flow is 10 times that of the Colorado River). Although its hydropower potential was recognized as early as the late 19th century, the immense size of the river and the lack of local markets capable of absorbing large amounts of power precluded the early development of hydroelectricity on the Columbia. With the onset of the Depression in the early 1930s, interest in using federal funds to finance construction of hydroelectric power facilities along the river received a major boost, especially after Franklin Roosevelt championed such projects during his 1932 election campaign. After Roosevelt assumed the presidency, politicians and businesspeople in Oregon and Washington began promoting federal funding for a series of dam projects on the Columbia that had been previously investigated by the Corps of Engineers.

The first of these projects to be authorized was the Bonneville Dam on the lower Columbia River about 40 miles east of Portland. The project included construction of (1) a 1,230-foot-long, straight-crested concrete overflow dam with a maximum hydraulic height of approximately 90 feet; (2) a powerhouse that ultimately housed 10 turbine-generator units with a capacity of more than 500,000 kilowatts; (3) a navigation lock with a hydraulic lift of more than 60 feet; (4) an extensive series of fishways (or fish ladders) to facilitate the upstream migration of salmon and other anadromous fish; and (5) administration buildings and residential structures. After extensive drilling the Corps of Engineers finally selected a site that straddled Bradford Island. The dam was erected across the main channel of the river, while the powerhouse was located across a secondary channel, or slough, between the island and the Oregon shore.

Construction began in 1934, and after three years of intensive effort, President Roosevelt returned to dedicate the completed dam in 1937. The power plant began operation a few years later and played a critical role in supplying electricity to the Portland region during World War II. The Bonneville Dam has been a major supplier to the Bonneville Power Administration, an independent federal agency that provides electricity to public and private users throughout the Northwest. In 1936 the state of Oregon also built a large fish hatchery immediately adjacent to the dam site. The dam-powerhouse-hatchery complex is now one of Oregon's most popular tourist attractions. NR, ASCE.

Bonneville Dam. Top: Generators in the powerhouse. Above: Fish ladder provided for spawning fish headed upstream. Below: Aerial view of the dam, with its powerhouse on the right.

CASCADE LOCKS

■ **Bridge of the Gods**
Across the Columbia River
On Bridge of the Gods Road,
adjacent to Interstate 84
Wauna Toll Bridge Company
1926

The Bridge of the Gods is a major example of a steel cantilever through truss spanning the second largest river in the United States. An interstate structure connecting Oregon and Washington, it was erected by the Wauna Toll Bridge Company rather than by any state or federal agency. The cantilever section of the bridge is 1,131 feet long with a clear span between piers of 706 feet. In the 1930s construction of the Bonneville Dam raised the level of the Columbia River, and the Bridge of the Gods had to be raised to facilitate river-borne transportation. Raising the bridge required the construction of approach spans that added more than 700 feet to its total length, but this work did not involve any significant alterations to the original cantilever structure. Its colorful name is taken from an Indian legend that describes a natural rock bridge at the site in ancient times.

Left: Cantilever truss forming the Bridge of the Gods. Right: Open-spandrel deck arches of the Rogue River Bridge.

GOLD BEACH

■ **Rogue River Bridge**
Across the Rogue River
On the Oregon Coast
Highway (U.S. Route 101)
Conde B. McCullough
1931

At the time of its completion, this was the largest bridge ever built by the Oregon highway department. The 1,898-foot-long structure includes seven spans of 230 feet each, with two ribbed, open-spandrel, reinforced-concrete deck arches. The main arches are supplemented by 18 concrete deck girder approach spans. Although the completed arches function as simple compression structures, they were built using prestressing construction techniques first developed by the French engineer Eugene Freyssinet in the early 1920s. The bridge was built as a centerpiece of the Oregon Coast Highway, and its designer, state bridge engineer Conde B. McCullough, consciously adopted a classical architectural motif to lend the structure beauty and grace. As well as providing Oregonians with an important intrastate transportation corridor, the Oregon Coast Highway also served as an attraction for America's motoring public visiting the Pacific Northwest. Handsome, solid structures such as the Rogue River Bridge gave confidence to tourists and reinforced the notion that travel in the state could be undertaken safely and efficiently. ASCE.

Caveman Bridge, with its three rainbow arches, c. 1935.

GRANTS PASS

Named the Caveman Bridge because of the proximity of Grants Pass to the Oregon Cave National Monument, this structure was built as part of the old Redwood Highway connecting southwestern Oregon with northern California. It is a regional variant of a reinforced-concrete rainbow arch and consists of three 150-foot-span, half-through arches. As documented in the Oregon Department of Transportation's recent inventory of historic bridges, it is one of only four structures of its type in the state. The Caveman Bridge was recently supplemented by a nearby girder span, so that the original multiarch bridge is required only to carry one-way traffic.

■ **Caveman Bridge**
Across the Rogue River
On U.S. Route 199
Conde B. McCullough
1931

LATOURELL

This graceful structure was built as part of the historic Columbia River Highway system. Incorporating more than 25 bridges into its design, the highway extends east from the Portland area through the rugged terrain of Columbia Gorge and serves as a major tourist attraction. The Shepperd's Dell Bridge is located on a section of the highway that encircles Crown Point, a major promontory rising more than 600 feet above the nearby Columbia River. The two-ribbed, open-spandrel, reinforced-concrete deck arch spans 100 feet and was designed under the auspices of the Oregon highway department. Construction of the Columbia River Highway began in 1913, the same year that the highway department also began operations, and the Shepperd's Dell Bridge is among the earliest surviving structures designed by the agency. It still carries highway traffic on the Crown Point Highway. ASCE.

■ **Shepperd's Dell Bridge**
Across Young Creek
In Shepperd's Dell State
Park, on Crown Point
Highway
K. R. Billner and
Samuel C. Lancaster
1914

Shepperd's Dell Bridge, one of the first bridges designed by the state highway department, c. 1920.

LINCOLN CITY

■ **Drift Creek Bridge**
Across Drift Creek
On Drift Creek County Road,
near U.S. Route 101
Lincoln County Public Works
Department
1914

Blessed with abundant forests, Oregon continued the tradition of wooden covered bridges well into the 20th century. In fact, a recent survey by the Oregon Department of Transportation documented 52 surviving covered highway bridges in the state built between 1914 and 1966. Most of these are Howe trusses constructed using standardized designs developed by the state highway department. The oldest covered bridge in Oregon is the 66-foot-long Drift Creek Bridge in coastal Lincoln County, only one and a half miles from the Pacific Ocean. In contrast to most surviving covered bridges in the state, it was designed by local rather than state officials. Now maintained by Lincoln County as a historic site, it no longer carries vehicular traffic. NR.

NORTH BEND

■ **Coos Bay (McCullough Memorial) Bridge**
Across Coos Bay
On the Oregon Coast
Highway (U.S. Route 101)
Conde B. McCullough
1936

Built as one of the last major links in the Oregon Coast Highway, the Coos Bay Bridge was posthumously dedicated to its designer and long-time Oregon state bridge engineer, Conde B. McCullough (1887–1946). Born in South Dakota, McCullough came to Oregon in 1916 to teach at Oregon State College. In 1919 he joined the Oregon highway department and during the next 27 years served as both state bridge engineer and assistant state highway engineer. McCullough clearly possessed a flair for visually pleasing engineering design, but he also appreciated the economic and political realities inherent in building 20th-century transportation systems. A prolific writer (his *Economics of Highway Bridge Types* was published in 1929), he also earned a law degree in order to be familiar with the legal implications of building and maintaining Oregon's highway system.

Designed just before McCullough went to Central America to work for a year planning and building the Inter-American Highway, the Coos Bay Bridge was the longest span in the state at the time of its completion. Built along with four other major Depression-era bridges on Oregon's coastal highway, it consists of a 1,709-foot-

Magnificent cantilever span across Coos Bay, perhaps Conde B. McCullough's finest design.

long steel through truss supplemented by 13 two-ribbed, open-spandrel, deck arch approach spans made of reinforced concrete. The entire structure is 5,305 feet long from shore to shore. Because Coos Bay is a commercially significant harbor, it was necessary to provide sufficient clearance for ocean-going ships. This requirement led to the selection of a steel cantilever structure in the middle of the bridge with a clear span of 793 feet between piers. After McCullough's death, the bridge was officially named in his honor in recognition of his efforts to improve Oregon's highway system.

OWHYEE

Oregon is frequently perceived as a lush, densely forested state so attractive that early settlers trekked thousands of miles along the Oregon Trail to reach it. Often forgotten is the fact that the eastern half of the state is a vast arid expanse of land that, in places, is among the least populated areas of the United States. In the late 1920s the Bureau of Reclamation began construction of a 417-foot-high, curved gravity dam on the Owhyee River, a tributary of the Snake River, as part of a project to increase irrigation in eastern Oregon. The dam was for a short period of time the highest water-impoundment structure in the world; the completion of the 726-foot-high Hoover Dam at Boulder City, Nev., in 1935 stole this bit of glory from the Owhyee Dam. It remains both an important part of eastern Oregon's economic base, however, and an imposing work of concrete construction in its own right.

■ **Owhyee Dam**
Across the Owhyee River
On Owhyee Dam Road, 15 miles south of State Route 201
U.S. Bureau of Reclamation
1932

Left: Owhyee Dam, a major precursor of the Hoover Dam. Right: David Steinman's St. John's Bridge.

PORTLAND

Located at the confluence of the Willamette and Columbia rivers in the northwest corner of the state, Portland is served by several river crossings. The most visually dramatic of these is the St. John's Bridge in the northeast section of the city. Designed by David B. Steinman, this suspension bridge has a main span of 1,207 feet, steel towers that each rise to a total height of 400 feet and a navigable clearance above sea level of more than 200 feet. It replaced a ferry that had been in continuous operation for 83 years. Aside from the 588-foot-span Crooked River Bridge (1963) built near the Cove Palisades State Park in Jefferson County, the St. John's Bridge is the only major suspension bridge in the state. At the time of its construc-

■ **St. John's Bridge**
Across the Willamette River
On U.S. Route 30 Bypass
David B. Steinman
1931

tion it was reputed to be the first suspension bridge in America built using prestressed rope strands rather than more conventional parallel wire cables. Because of the Gothic arches that adorn its towers, the St. John's Bridge is a favorite of bridge lovers in the Portland area.

UMATILLA

■ **Three-Mile Falls Dam**
Across the Umatilla River
2 miles south of U.S.
Route 730
U.S. Reclamation Service
1914

Before the Reclamation Service changed its name and became the Bureau of Reclamation in the early 1920s, it built only one reinforced-concrete, multiple-arch dam — the Three-Mile Falls Dam on the Umatilla River in northeastern Oregon. Rising 24 feet above its foundation and stretching 915 feet along the crest, the dam is located only about three miles south of the Columbia River. It is used to divert irrigation water for bottom land adjacent to the larger stream. The structure currently functions as part of a larger system controlled by the McKay Dam about 50 miles upstream on the Umatilla River.

■ ■ ■ ■ ■ ■ ■ UTAH ■ ■ ■ ■ ■ ■ ■

CHARLESTON

■ **Deer Creek Dam**
Across the Provo River
On U.S. Route 189
U.S. Bureau of Reclamation
1941

The Provo River is one of central Utah's most important large waterways. Flowing out of the eastern Uinta Mountains near the Wyoming border, it travels southwest for almost 150 miles before emptying into the freshwater Utah Lake, near Provo. Along the way it passes through the Wasatch Mountains in a deep, long gorge known as Provo Canyon. Mormon settlers near Provo first used water from the Provo River for irrigation in the early 1850s, but there was no large reservoir on the river to store spring floods until construction of the Deer Creek Dam in the upper Provo Canyon. Built by the Bureau of Reclamation using federal funds, the dam substantially increased the amount of water available for farmers in the region surrounding Provo. It is a 235-foot-high, 1,304-foot-long rockfill dam that carries Route 189 across the crest. The Deer Creek Dam has subsequently developed into a major recreation site for water sports.

DUTCH JOHN

■ **Flaming Gorge Dam**
Across the Green River
Adjacent to U.S. Route 191
U.S. Bureau of Reclamation
1964

The Green River drains a large portion of western Wyoming and constitutes a major tributary of the Colorado River. As part of its plans to develop fully the hydroelectric power potential of the river's watershed, the Bureau of Reclamation had long planned to build a major dam at the Flaming Gorge site in eastern Utah. The project engendered considerable opposition from environmental groups, but in 1964 the bureau completed construction of the 502-foot-high concrete arch dam. The Flaming Gorge Dam represents the last major dam built on the main stem of the Colorado River, and it continues to function as a major source of electricity for the southwestern United States.

Left: Artist's rendering of the Flaming Gorge Dam. Below: Rockfill Echo Dam, with its spillway on the right.

ECHO

The Echo Dam is a 158-foot-high, 1,887-foot-long earth-fill structure built to store water for irrigation in the Weber River Valley near Ogden. Studies of the dam site began as early as 1905, but financial authorization for the project was not granted until the late 1920s. Construction of the dam, still in active use, also involved costly realignment of the Lincoln Highway and a heavily traveled section of the Union Pacific Railroad.

■ **Echo Dam**
Across the Weber River
Adjacent to Interstate 80
U.S. Bureau of Reclamation
1931

FRUITLAND

When Mormon pioneers came to Utah in the late 1840s, they concentrated their settlements around the Great Salt Lake and other parts of the Great Basin in the central and southern parts of the territory. Brigham Young expressed little interest in eastern Utah, which was part of the Colorado River watershed. As a result, much of this region was left to Indian settlement. In the early 20th century the Reclamation Service began searching for a major irrigation project to build in the state, and attention quickly focused on this previously under-developed area. But rather than build a project that would facilitate completely new irrigation activity, the Reclamation Service chose to support an already pro-ductive agricultural area. Known as the Strawberry Project, the service's plan involved building a dam on the Strawberry River, a tributary of the Colorado River. Water would be diverted into a tunnel extending through the Wasatch Mountains, where it would empty into Diamond Creek, a tributary of Utah Lake.

■ **Strawberry Dam**
Across the Strawberry River
Adjacent to U.S. Route 40
U.S. Reclamation Service
1913

Left: Construction of the Strawberry Dam, c. 1910.
Right: Completed Strawberry Dam, c. 1915.

The Strawberry Project represented one of the first major interbasin transfers of water in the American West in that water that would normally flow into the Pacific Ocean was transferred for the use of farmers in the Great Basin. The Strawberry Dam, a 72-foot-high, 490-foot-long earthfill structure, is a key component of this still-active system. NR.

OREM

■ **Provo River Railroad Bridge**
Across the Provo River
Adjacent to U.S. Route 189,
at the mouth of Provo Canyon
Union Bridge Company
1884; reerected 1909 and 1919

In 1884 the Denver, Rio Grande and Western Railroad built a three-span Pratt truss bridge across the Green River in eastern Utah. After 1900 the bridge was dismantled and one of the spans shortened to 82 feet. The structure was then reerected over the Price River near Wellington. In 1919 the pin-connected truss made a final journey to its present site across the Provo River, near Orem. The structure carried traffic on the railroad's Provo Canyon line to Heber until 1969, when it was abandoned. Recently, it found new life as a component of the Heber Creeper railroad, one of Utah's most popular tourist attractions. Technologically, the Provo River Railroad Bridge is significant in demonstrating the great adaptability of metal truss technology to a variety of sites.

Pratt truss railroad bridge across the Provo River, its third location.

Above: Ogden-Lucin Cutoff Trestle in 1971. Left: Construction crew working on the trestle in 1903.

PROMONTORY POINT

When the Central Pacific Railroad built the western section of the first transcontinental railroad in the 1860s, it did not always have the ability to select an ideal location for the right-of-way. In northern Utah, for example, it was forced to take a rather circuitous route to avoid the Great Salt Lake. When E. H. Harriman took over the Southern Pacific Railroad, the successor to the Central Pacific, shortly after the turn of the century, he started a major program to eliminate uneconomical sections of track and realign the railroad for more efficient service. To this end he authorized building a new right-of-way between Ogden and Lucin that would be 43 miles shorter than the original line and involved erecting a wooden trestle more than 11 miles long across the Great Salt Lake. Construction began in 1902, and by March 1904 trains were using the structure for regular service. The trestle is still in place, but the unprecedented rise of the water level of the Great Salt Lake in the 1980s threatens its survival. NR.

■ **Ogden-Lucin Cutoff Trestle**
Across the Great Salt Lake
Along the Southern Pacific
Railroad right-of-way,
between Promontory and
Lakeside
William Hood
1904

Mountain Dell Dam. Above: Upstream side of the dam in 1925 after its height was raised 40 feet. Above right: Construction of the downstream side, 1925. Right: Buttresses and strut-tie beams of the downstream side as they appear today.

SALT LAKE CITY

■ **Mountain Dell Dam**
Across Parley's Creek
Adjacent to Interstate 80
John S. Eastwood
1917, 1925

In 1915 Salt Lake City requested contractors to bid on three different designs for a new, 150-foot-high storage dam in Parley's Canyon. Prepared under the direction of Sylvester Q. Cannon, the city engineer, these designs included a concrete curved gravity dam, a reinforced-concrete Ambursen flat-slab dam and an Eastwood multiple-arch dam. More than 10 contractors submitted bids for the job, and the results provided a striking demonstration of the economic advantages of multiple-arch designs. Whereas the low bids for the gravity and Ambursen designs were well over $200,000, the multiple-arch design came in at less than $140,000. Salt Lake City subsequently decided to build a multiple-arch structure in Parley's Canyon; it was built to an initial height of 110 feet in 1917 and raised to 150 feet in 1925. Cannon advocated the design submitted by John S. East-wood because of its low cost and because he wanted a completely safe structure on the relatively weak foundations of the Mountain Dell site. Cannon later became a member of the Mormon Church's Quorum of Twelve, the religion's highest council. His faith in the safety of Eastwood's Mountain Dell Dam was not misplaced, and the dam remains an important part of the municipal water supply system. NR.

■ ■ ■ ■ ■ ■   WASHINGTON   ■ ■ ■ ■ ■ ■

COULEE DAM

The Grand Coulee is the name for a natural geological phenomenon formed when, in prehistoric times, the Columbia River left its original channel and carved a huge gash in the earth's crust. This depression was 600 to 800 feet deep, two to six miles wide and 50 miles long. The Grand Coulee Dam is located at the lower end of this natural basin. Beginning in the late 19th century, settlers in the Pacific Northwest realized the desirability of building a major storage dam on the Columbia and recognized the attractiveness of a site near the Grand Coulee. But the cost of such a project was so great that it remained

■ **Grand Coulee Dam**
Across the Columbia River
Adjacent to State Route 155
U.S. Bureau of Reclamation
1942

Grand Coulee Dam. Above: Downstream site of the dam before construction had begun. Left: Construction under way, c. 1940. Below: Dam in operation, c. 1945.

little more than a pipe dream. After all, the Columbia is the second-largest river in the United States (only the lower Mississippi exceeds it), and any storage dam designed to impound its flow required tens of millions of dollars to build.

The opportunity to turn this dream into reality came with Franklin Roosevelt's election as president in late 1932. Roosevelt saw the project as an ideal means of bringing his vision of economic revitalization to the Pacific Northwest through massive financing for public works. The project appealed to him also because of its high visibility and the undeniable satisfaction that could come with taming one of the world's greatest rivers for the benefit of humanity. Federal authorization for the project was given in 1933, and in the summer of 1934 Roosevelt personally visited the dam site to oversee ceremonies marking the start of construction.

The Grand Coulee Dam is a 550-foot high, 4,173-foot-long concrete gravity dam that is among the most massive structures ever built in the world. Water stored behind the dam serves two primary purposes. First, much of it is released through the dam's hydroelectric power plant and is used to generate vast quantities of power. This power is transmitted throughout the Pacific Northwest region. Second, some of the power is used at the dam site itself to pump water up 280 feet to the Grand Coulee proper, forming a secondary reservoir that supplies irrigation water to an area of several hundred square miles in central Washington. Thus, strange as it may seem, the Grand Coulee Dam does not actually form the reservoir that inundates the Grand Coulee itself.

Construction of the dam took eight years and was not completed until 1942. Initially, there were fears that no use would be found for the huge quantities of electric power generated at the dam. During World War II, however, the dam proved critically important to activities at the Hanford Reservation, where much work on the first atomic bombs took place. The dam also provided the impetus for building many aircraft-manufacturing facilities in the region. Since its completion, other dams have been built across the Columbia River, but none of them comes close to rivaling the grandeur of the Grand Coulee Dam. NR.

DIABLO

■ Diablo Dam
Across the Skagit River
Adjacent to State Route 20
Constant Angle Arch
Dam Company
1929

As part of its system to develop hydroelectric power on the Skagit River in northern Washington, the city of Seattle began construction of the Diablo Dam in late 1927. It was designed by the Constant Angle Arch Dam Company, a firm established by Lars Jorgensen, engineer of the Salmon Creek Dam (1914) near Juneau, Alaska. The structure is more than 400 feet high from the deepest foundation to the top and has a maximum thickness of about 135 feet. The conical shape of the constant-angle arch design was particularly well suited to the narrow granite canyon. The dam provides storage for 90,000 acre-feet of water that can be used in a series of power

Diablo Dam, built to supply Seattle with hydroelectric power.

plants built along the river below the dam. Located 100 miles northeast of Seattle, the Diablo Dam continues today as an important part of the municipality's publicly owned electric power system.

LONGVIEW

Built by the privately financed Columbia River Longview Company as a toll bridge, the Longview Bridge primarily consists of a 2,722-foot-long steel cantilever truss that rises 195 feet above the river level. The main span of the cantilever stretches 1,200 feet between piers and provides adequate clearance for the extensive shipping that plies the lower Columbia River. The bridge is noteworthy because it was designed by Joseph B. Strauss, chief engineer for the Golden Gate Bridge. It was the longest span cantilever bridge in the world at the time of completion. Today, the structure is owned by the Washington State Department of Transportation and remains open to highway traffic. NR.

■ **Longview (Lewis and Clark) Bridge**
Across the Columbia River
On State Route 433
Joseph B. Strauss
1930

Longview Bridge, designed by the same engineer responsible for the Golden Gate Bridge.

LYONS FERRY

In 1927 county authorities built a 1,636-foot-long steel cantilever through truss across the Columbia River at the city of Vantage. Following construction of the nearby Wanapum Dam in the early 1960s, the bridge was replaced by a new structure located at a higher elevation above the river. The steel members from the 1927 bridge were subsequently disassembled and, in 1968, reerected at Lyons Ferry across the Snake River. Aside from the addition of four new concrete girder approach spans, the rebuilt structure is almost identical to the original span. The reerection of this bridge provides yet another striking demonstration of the flexibility of steel truss technology.

■ **Lyons Ferry Bridge**
Across the Snake River
On State Route 262
Washington Department of Transportation
1927; reerected 1968

Above: Pasco-Kennewick cantilever span, c. 1940. Right: Old structure, now overshadowed by a new cable-stayed suspension bridge. In the background is a railroad bridge.

PASCO

■ **Pasco-Kennewick Bridge**
Across the Columbia River
Adjacent to U.S. Route 12
Union Bridge Company
1922

Built to replace a ferry capable of carrying only six cars per trip, the Pasco-Kennewick Bridge is a 3,300-foot-long steel structure that provided the first permanent connection between Pasco and Kennewick. The main part of the structure consists of a cantilevered through truss with a clear span of 432 feet and two adjacent 234-foot-long Warren through trusses. With its completion in 1922, the bridge became a key component in the transcontinental Yellowstone Trail. In the late 1970s the historic cantilever span was replaced by a new cable-stayed suspension bridge built directly adjacent to it. As the new bridge carries all highway traffic at the site, the 1922 bridge was scheduled for demolition. However, persistent community activism thus far has preserved the old bridge in place (see page 68). NR.

SEATTLE

■ **Lacey V. Murrow (Lake Washington) Floating Bridge**
Across Lake Washington
On Interstate 90
Charles Andrew and Clark Elkridge
1940

To provide a direct highway route from Seattle east through the Cascade Mountains, the Washington Toll Bridge Authority built this floating structure across Lake Washington. Because the lake has an average depth of more than 150 feet, constructing piers for the 7,800-foot-long crossing would have been economically impossible. Instead, the authority selected a design in which 25 reinforced-concrete floating pontoons are used to support the highway deck. Each of the pontoons is 350 feet

long, 60 feet wide and 14 feet deep. Internally, they are divided into a series of watertight compartments so that a leak in one part of the pontoon cannot flood the entire structure. The floating section of the bridge is supplemented by two steel Warren trusses, a steel arch and three reinforced-concrete girder spans that allow clearance for the passage of small ships. Named after the state highway director in office during the bridge's construction, the Lacey V. Morrow Floating Bridge is unlike any other reinforced-concrete structure in the United States. Because of the great depth of Lake Washington and the fact that it is devoid of currents, drift or ice, the large pontoon design was uniquely suited to the site. It has undergone renovation and refurbishment during the past 45 years, but, in general, it retains the integrity of its original design. The bridge continues to carry heavy volumes of highway traffic. NR.

Lacey V. Murrow Floating Bridge shortly after completion in 1940.

TAHOLAH

The Chow Chow Bridge is a rather remarkable wood and steel cable-stayed suspension bridge built by the Aloha Lumber Company as part of its lumber operations in the Quinault Indian Reservation. Designed by Frank Milward, the company's logging superintendent and a man with no previous experience in bridge construction, the Chow Chow Bridge and a nearby bridge of similar design (no longer extant) were examples of homegrown structural engineering. The Chow Chow Bridge is supported on cedar towers and has a clear span of approximately

■ **Chow Chow Bridge**
Across the Quinault River
In the Quinault Indian
Reservation, near State
Route 109
Frank Milward
1952

Chow Chow Bridge, an example of homegrown structural technology.

Chow Chow Bridge, in the Quinault Indian Reservation, showing king-post trusses that support the traffic deck.

190 feet. The traffic deck consists of four king-post trusses that are connected to the suspension cables. Although no longer used for lumber trucks, the bridge still carries local traffic in the reservation. Visitors should first check with local tribal officials for permission to visit the bridge. NR.

■ ■ ■ ■ ■ ■ ■   WYOMING   ■ ■ ■ ■ ■ ■ ■

ALCOVA

■ **Pathfinder Dam**
Across the North Platte River
7 miles south of State Route 220
U.S. Reclamation Service
1910

Masonry arch of the Pathfinder Dam in southeastern Wyoming, 1916.

Following passage of the federal Reclamation Act in 1902, the Reclamation Service set out to find viable irrigation projects in regions west of the 100th meridian. For political and legal reasons the Reclamation Service needed to build projects in as many states as possible. The North Platte River flows through much of Wyoming and Nebraska, and construction of a large storage dam

on the river offered an opportunity to develop irrigation projects in both states. With this in mind, the Reclamation Service decided to build the Pathfinder Dam on the North Platte as a way to please farmers and politicians in the two states. Construction began in late 1905, and work continued without incident until June 1910. The structure is a 218-foot high, masonry arch dam with a maximum base width of 94 feet, considerably less than the width a gravity design would require. Since its completion, the Pathfinder Dam has been supplemented by several other storage dams on the North Platte, but it remains a key part of the regional water supply system. It is named after Gen. John C. Fremont, a famous 19th-century western explorer known as the Pathfinder. NR.

CODY

Flowing out of the mountains that border the eastern edge of Yellowstone National Park, the Shoshone River passes through a dramatically narrow canyon about 10 miles west of Cody. Only 70 feet wide at the bottom and 200 feet wide at a height 200 feet above the stream bed, the canyon offered an ideal location for a large storage dam. Situated so that its reservoir could facilitate irrigation in a large area of north-central Wyoming, it was selected by the Reclamation Service as the site of one of its first dams. Because bedrock in the canyon consists of hard granite, the service felt secure building a thin-arch dam 328 feet wide with a maximum height of 108 feet. The Buffalo Bill Dam is among the thinnest ever built by the Reclamation Service or its successor, the Bureau of Reclamation. Construction of the concrete structure began in 1905 and topped out five years later. The dam's original name came from the river, but after World War II it was changed in honor of Cody's most famous citizen. NR, ASCE.

■ **Buffalo Bill (Shoshone) Dam**
Across the Shoshone River 1 mile south of U.S. Route 20
U.S. Reclamation Service 1910

Left: Upstream face of the Buffalo Bill Dam, an arch dam. Right: Downstream side, showing the narrow canyon at the dam site.

DOUGLAS

■ La Prele (Douglas) Dam
Across La Prele Creek
3 miles east of State
Route 91
Ambursen Hydraulic
Construction Company
1909

The Boston-based Ambursen Hydraulic Construction Company first developed its flat-slab buttress dam designs in New England, but the company soon sought to expand its market westward. In 1908 the firm received the design commission for a 135-foot-high, 360-foot-long, reinforced-concrete flat-slab buttress dam across La Prele Creek in east-central Wyoming. La Prele Creek is a small tributary of the North Platte River with sufficient flow to support several thousand acres of irrigated cropland. In 1907 the La Prele Ditch and Reservoir Company initiated work on a major irrigation system for the Douglas area that would rely on a large storage reservoir on La Prele Creek. To meet this need the Ambursen Company undertook the design and construction of the world's first reinforced-concrete buttress dam to exceed 100 feet in height. The La Prele Dam (also called the Douglas Dam after the nearby town) has buttresses spaced 18 feet apart center-to-center. Aside from their splayed foundations, the buttresses have a maximum thickness of 50 inches. The upstream face of the dam is 12 inches thick at the crest and increases to 54 inches at the deepest section. The face is inclined upstream at an angle of approximately 40 degrees, which gives the structure great stability under hydrostatic loads. In the late 1970s the dam underwent some minor rehabilitation work to increase the spillway capacity, but, other than this, it retains most of its original design integrity. It remains an important component of eastern Wyoming's agricultural economy.

FORT LARAMIE

■ Fort Laramie Bowstring Arch Truss Bridge
Across the North Platte
River
In Fort Laramie National
Historic Site, near State
Route 160
King Iron Bridge and
Manufacturing Company
1875

Beginning in the 1830s, pioneer settlers blazed the Oregon Trail from Independence, Mo., to the Pacific Northwest. The North Platte River Valley served as a major conduit for this transportation route, and Fort Laramie developed as an important way station for emigrants headed west. Following completion of the transcontinental railroad in 1869, the importance of the Oregon Trail waned, but Fort Laramie remained significant as Indians were resettled and Anglo-American homesteading and mining developed in the region. In 1874 Congress authorized the army to construct a

Three-span bowstring arch truss at Fort Laramie, among the longest structures of its type to survive in the United States.

Details of the abutment of the Fort Laramie Bridge.

permanent bridge across the North Platte at Fort Laramie for public use. The government soon accepted plans developed by the King Iron Bridge and Manufacturing Company, and in early 1875 components for a three-span, 400-foot-long bowstring arch truss were shipped to the site. Construction was completed in December, and the bridge remained under army control until the last regular garrison departed in 1890. After the structure passed through the hands of the U.S. Department of the Interior, it was taken over by local authorities in 1894 and provided highway service until 1958. Shortly afterward it reverted to federal control and is now maintained as part of the Fort Laramie National Historic Site. NR.

MORTON

With construction of the Wind River Diversion Dam as part of a local irrigation project, the Bureau of Reclamation provided Wyoming's highway department with an opportunity to build a bridge over the Wind River at a relatively reasonable cost. The combination dam-bridge is more than 600 feet long and includes eight Warren pony trusses. Taken alone, neither the Warren pony trusses nor the diversion dam is particularly remarkable in size. However, the structure supports one of the longest bridges in Wyoming and represents an interesting melding of bridge and dam technologies. NR.

■ **Wind River Diversion Dam-Bridge**
Across the Wind River
On County Road CN 10-24, 1 mile north of U.S. Route 26/287, 9 miles west of Morton
U.S. Bureau of Reclamation (dam)
Wyoming Highway Department (bridge)
1925

Structure across the Wind River, used both to divert water for irrigation and support an eight-span Pratt pony truss.

■ ■ ■ ■ ■ ■ ■ ■ ■ ■ ■ ■ ■ ■ ■ ■ ■

EPILOGUE:
THE ONES THAT GOT AWAY

Thousands of American bridges and dams have come and gone. Some of these collapsed because of design flaws and human error or because of extreme natural conditions. Others were razed or replaced to make way for improved transportation or water supply systems. Many of the structures included in this book could similarly disappear from the American landscape in the years ahead. What follows here are the stories of some of the more notable American bridges and dams destroyed or demolished during the past 150 years or scheduled for imminent replacement. The examples were selected for a variety of reasons — their notoriety as well as the fact that circumstances surrounding their demise often represent problems faced by still-surviving structures.

Opposite: Remains of the St. Francis Dam north of Los Angeles after its collapse in 1928.

BRIDGES

In 1808 James Finley received a patent for a suspension bridge using iron link chains for the main tension members and wooden towers. During the next decade Finley built numerous bridges using his design and also licensed others, including John Templeman, to use the patent. Among the most famous of these spans was a suspension bridge across the Merrimack River near Newburyport, Mass. A heavily loaded ox-drawn wagon damaged the bridge in 1827, but the span was rebuilt closely following the original design. This structure survived until 1913, when it was completely replaced by a bridge using wire suspension cables and reinforced-concrete towers. All that remains of the early 19th-century bridge are some of the foundations.

■ **Newburyport (Essex-Merrimack) Suspension Bridge**
Across the Merrimack River at Deer Island
Newburyport, Mass.
John Templeman
1810–1913

Having a clear span or more than 340 feet, the Colossus covered bridge was the most stunning and visually compelling engineering structure built in the early United States. The all-wood design represented the pinnacle of America's "Wooden Age" and provided a major crossing of the Schuylkill River in the heart of Philadelphia. The fame of the Colossus also gave a major boost to its builder's covered-bridge construction business. In 1838 the Colossus suffered a fate that has befallen many other wooden bridges: it burned. Although it existed for only 25 years, the Colossus must still be considered one of the greatest of all American bridges.

■ **Colossus Bridge**
Across the Schuylkill River
Philadelphia, Pa.
Louis Wernwag, builder
1813–38

For many years this 97-foot-long, cast-iron and wrought-iron Fink through truss was among the oldest surviving metal truss bridges in the United States. Carefully maintained by Hunterdon County's highway department, it carried local traffic for more than 120 years. Unfortunately, in September 1978, only a few months before it was scheduled to be bypassed by a new bridge and preserved for pedestrian use as part of a county park,

■ **Fink Truss Bridge**
Across the South Branch of the Raritan River
Hamden, N.J.
Trenton Locomotive and Machine Manufacturing Company
1857–1978

the Fink through truss was hit by an automobile. The cast-iron portal compression member shattered, and the entire span collapsed into a tangled mess. Despite an effort to reconstruct the bridge using newly cast members, the Hamden bridge today remains accessible to historians only through archival photos and measurements made before its unanticipated demise.

■ **Ashtabula Bridge**
Across the Ashtabula River
Ashtabula, Ohio
Amasa Stone
1865–76

Although the extent of late 19th-century bridge failures has been grossly exaggerated, a number of tragic collapses gained widespread notoriety. The most famous of these was the failure of a 157-foot-long, all-metal Howe deck truss in Ashtabula, Ohio, on the night of December 29, 1876. The bridge was built under the direction of Amasa Stone, president of the Lake Shore and Michigan Southern Railroad, without any guidance from an experienced engineer. With the rapid expansion of America's railroad system in the mid-19th century, bridges were designed by a variety of craftsmen, engineers, bridge companies and, at times, to speed up construction, even railroad executives. The disaster occurred during a severe snowstorm (which probably helped precipitate the collapse because of ice buildup on the iron structure) and sent an 11-car train plunging into the Ashtabula River. More than 80 persons died in the ensuing inferno. The tragic accident attracted national attention and prompted the engineering profession to castigate any bridge construction that was not handled by experienced bridge engineers. Because the cause of the collapse was blamed largely on the failure of some cast-iron castings, the use of cast iron in future bridge design was almost completely abandoned.

■ **Poughkeepsie
Railroad Bridge**
Across the Hudson River
Poughkeepsie, N.Y.
J. F. O'Rourke,
P. P. Dickenson and
A. B. Paine
1888

In the 1870s the Hartford and Connecticut Western Railway received a charter to build a major railroad bridge over the Hudson River at Poughkeepsie. After a delay of many years, work finally began on building a 6,767-foot-long, multispan steel cantilever deck truss in 1886. Completed two years later, the massive Poughkeepsie Railroad Bridge stood as one of the largest operating bridges in the United States for the next 86 years. In 1974 Conrail closed it to traffic following a fire that damaged the deck structure. Since that time all of the railroad rails

City of Poughkeepsie, with the railroad bridge looming in the background.

Left: Original Newburyport Bridge, with its wooden towers and chain link cables. Below: View of Louis Wernwag's Collossus Bridge in Philadelphia, printed before its destruction by fire in 1838.

Left: Fink Truss Bridge, Hamden, N.J., before its collapse in 1978. Below: Remains of the Ashtabula Bridge in northeastern Ohio after the disaster of 1876.

Poughkeepsie's bridge as it looked at the turn of the century.

have been removed from both approaches to the bridge, and it is estimated that putting the bridge back into operating condition could cost upwards of $25 million. At present the bridge is owned by Railway Management Associates, a group that purchased it from Conrail for the grand sum of one dollar.

Perhaps it may be premature to consider the Poughkeepsie Railroad Bridge "one that got away," but many preservationists are pessimistic that long-term economically viable means of preserving the bridge can be found. With luck, they may be proven wrong. But the question still remains, What can one do with a damaged railroad bridge that will cost many million dollars to repair and really isn't needed as part of any significant transportation system? If you have any good ideas, please get in touch with the city authorities in Poughkeepsie.

■ **Smith Avenue High Bridge**
Across the Mississippi River
Minneapolis, Minn.
Keystone Bridge Company
1889–1985

For almost a century, this 2,770-foot-long, wrought-iron bridge carried highway traffic on Smith Avenue over the Mississippi River. But in the summer of 1984 the Minnesota transportation department completely closed the pin-connected, multispan Warren deck truss bridge to all traffic because of concerns about its structural integrity. As the longest surviving 19th-century bridge in the Twin Cities, the bridge had many admirers. But the sudden closure of the span prompted residents to demand rapid construction of a new bridge. The possibility of repairing the existing structure was dismissed by city and state highway engineers as infeasible, especially because of concern that the wrought-iron pins used to hold together the truss were severely deteriorated. Thus, on February 25, 1985, the bridge was dynamited into the Mississippi River as a crowd of 25,000 persons looked on. Ironically, divers later salvaged the wrought-iron pins from the wreckage, and, as reported in the Society for Industrial Archeology newsletter, they were found to be in excellent condition. Unfortunately, the circumstances surrounding the destruction of the Smith Avenue High Bridge are not unusual. If nothing else, they reflect the difficulty of trying to prove an old bridge to be safe in the face of conservative engineering opinions and the massive amounts of federal funding now available for highway bridge replacement.

After being closed to traffic for more than a decade because it was thought to be structurally inadequate, the stunning Bellows Falls Arch Bridge dramatically fought off an onslaught of explosives before it succumbed in 1982. Two days and five blasts were needed finally to level the 540-foot-long span, the longest through arch in the United States when it was built. After a 1971 study found the deck too deteriorated for vehicular use, plans were made for a new bridge. A later analysis funded by the National Trust and the Society for Industrial Archeology found, instead, that the parabolic-arch structure could be rehabilitated for $1.1 million, about the cost of demolition. But New Hampshire, the major owner of the bridge spanning the river between Bellows Falls and North Walpole, N.H., was not interested in helping to get the historic span back into service. The state told the two communities to pick up the tab if they wanted the bridge saved. Having been led to believe that it was not salvageable, and not being in a position to raise the

■ **Bellows Falls Arch Bridge**
Across the Connecticut River
Bellows Falls, Vt.
J. R. Worcester
1905–82

Bellows Falls Arch Bridge. Below: In the 1960s, looking toward the New Hampshire side. Bottom: Withstanding the demolition squad's initial blasts. The bridge succumbed to heavier charges a short time later.

needed revenues, the towns declined the offer. Although the bridge was determined eligible for the National Register of Historic Places, neither the federal review process nor local interest was sufficient to spare the landmark. Wrote the Rutland (Vt.) *Herald* after the bridge's valiant final battle, "The bridge seemed to mock the engineers from both Vermont and New Hampshire who said the bridge had outlived its usefulness and was a serious public danger."

■ **Point Pleasant
(Silver) Bridge**
Across the Ohio River
Point Pleasant, Ohio
J. E. Greiner Company
1928–67

The growth of America's highway system in the 1920s prompted the construction of many new long-span crossings of major rivers. In 1927 the J. E. Greiner Company of Baltimore began constructing a pair of similar suspension bridges over the Ohio River between Ohio and West Virginia. Located at Point Pleasant and St. Marys, W. Va., the two spans were distinguished by the use of eyebar link chains (instead of wire cables) for the tension members. The Point Pleasant Bridge had a clear span of 700 feet between towers and was known as the Silver Bridge because of its paint color. The structure served regional transportation needs without incident until, without warning, on December 15, 1967, the Point Pleasant Bridge collapsed and sent 46 people to their death. After extensive study and analysis of the wreckage, the failure was blamed on stress fatigue. This fatigue,

Deck and tower of the ill-fated Point Pleasant Bridge, c. 1940.

which occurred from years of repeated heavy loads, was found to have caused hairline cracks in the steel pins holding together the eyebars. After the Point Pleasant disaster, its sister span at St. Marys was demolished as a precautionary move. The failure of the Point Pleasant Bridge prompted enactment of the first federal legislation to fund the replacement of unsafe highway bridges. Even today, the disaster is still used by the highway industry to help justify the need for a massive federal bridge and replacement program.

The Alsea Bay Bridge is one of the most picturesque highway structures in Oregon, a state renowned for its beautiful bridges. Yet, despite widespread aesthetic appreciation of the Alsea Bay Bridge among engineers and the general public, the massive structure is scheduled for replacement in 1989. The problem with the 3,011-foot-long, reinforced-concrete, combination deck and through arch highway crossing is not solely the condition of the superstructure above the waterline. It is also the deteriorated condition of the foundations, which, according to studies by the Oregon transportation department, are verging on collapse. This condition is exacerbated by the marine environment of the site, unfortunately precluding any possibility that the huge structure can be economically rehabilitated for continued use.

■ **Alsea Bay Bridge**
Across Alsea Bay
Waldport, Ore.
Conde B. McCullough
1936

Alsea Bay Bridge, one of Conde B. McCullough's most graceful reinforced-concrete designs.

On July 1, 1940, the Tacoma Narrows Bridge opened for traffic and became one of the most dramatic-looking suspension bridges in the world. Built with a clear span of 2,800 feet between towers, it was the third longest bridge in the world at the time of construction. The design also was noteworthy because the plate-girder deck was only eight feet deep. Located in a windy part of the Pacific Northwest, the bridge soon became known as Galloping Gertie because the traffic deck would often undulate with an intensity sufficient to induce seasickness. The oscillations of the deck caused great concern among the engineers responsible for its design, but before they could do anything to stop the movement,

■ **Tacoma Narrows Bridge**
Across the Tacoma Narrows
Tacoma, Wash.
Leon Moisseiff
1940

Right: Deck of the original Tacoma Narrows Bridge undulating before its collapse in 1940. Below: Earthfill South Fork Dam after being overtopped in 1889.

the bridge devised its own method of solving the problem. On November 7, 1940, a strong windstorm began to oscillate the bridge with severe intensity. The twisting and writhing of the deck progressively increased until the structure ripped itself apart and collapsed into the water. The destruction of the Tacoma Narrows Bridge caused civil engineers to recognize the importance of bracing suspension bridges against wind stresses and building relatively deep deck trusses that were less susceptible to oscillation under wind loadings. After the Tacoma Narrows failure, suspension bridges would never again be built with such thin proportions for the traffic deck.

DAMS

■ **South Fork Dam**
Across the South Fork of the Conemaugh River
Johnstown, Pa.
Pennsylvania State Canal Company
1853–89

To increase the water supply for the Pennsylvania State Canal, canal authorities began building a storage dam on the South Fork of the Conemaugh River in the late 1830s. Construction proceeded fitfully and did not conclude until the early 1850s. Shortly afterward the canal itself was rendered obsolete by competition from the Pennsylvania Railroad, and by the early 1880s the South Fork Dam had come under the control of Pittsburgh busi-

nessmen who used the reservoir for recreational purposes. The South Fork Hunting and Fishing Club operated the 72-foot-high, 800-foot-long earthfill dam until May 31, 1889. Then a severe rainstorm drenched western Pennsylvania and caused the flow in the South Fork of the Conemaugh River to exceed the dam's spillway capacity. As a result, water overtopped the dam and caused the earthen structure to erode rapidly. Within a few minutes after overtopping, a deluge of water descended on the city of Johnstown and caused enormous damage. Approximately 2,000 people died, making it one of the most devastating disasters in American history. The carnage associated with the Johnstown Flood became a major chapter in American folklore and demonstrated the terrible consequences that can result from the overtopping of an earthfill dam.

The Santa Ana River is the largest waterway in southern California and is the source of water for irrigated land in the San Bernardino region. Early farmers in the area soon discovered an ideal reservoir site along Bear Creek, a major tributary of this river. Located at an elevation of more than 6,000 feet, this site presented a special challenge to the Bear Valley Mutual Water Company because of its remoteness from major transportation routes. To reduce construction costs, the company chose to build a 64-foot-high, thin-arch masonry dam that required much less material than a comparable gravity dam. This design minimized the amount of cement and stone to be transported to the site, and it also helped minimize labor costs. The daring design functioned safely for more than 25 years before the company

■ **Bear Valley Dam**
Across Bear Creek
Big Bear City, Calif.
F. E. Brown
1884–1911

F. E. Brown's Bear Valley Dam, the world's most daring arch dam for more than 25 years.

expanded the size of its reservoir, known as Big Bear Lake, by building a taller reinforced-concrete, multiple-arch dam a short distance downstream. This new dam, built by John S. Eastwood in 1910–11, completely inundated the 1884 dam except during periods of low water. In recent years a thriving resort community has developed around Big Bear Lake and obtained legal control over the elevation of the reservoir. The local tourist industry is dependent on a full reservoir, so it is unlikely that the lake will ever again drop to levels that will allow the 1884 dam to be visible. Intense planning also is under way to either inundate the 1911 multiple-arch dam behind a new concrete gravity structure or drastically alter its original design.

■ **Austin Dam**
Across the Colorado River
Austin, Tex.
Joseph P. Frizell and
J. T. Fanning
1892–1900

In the early 1890s the city of Austin erected a 1,091-foot-long, 68-high-foot masonry gravity overflow dam across the Colorado River for the production of hydroelectric power. Located about two and one-half miles upstream from the city, the dam operated successfully for almost 10 years until heavy flooding hit the Colorado River in the spring of 1900. On April 7, 1900, the dam was overtopped to a depth of more than 11 feet, which resulted in a 500-foot-long section of the structure sliding 60 feet downstream, thus causing the release of all water stored in the reservoir. The failure of the Austin Dam highlighted the danger posed to gravity overflow dams by excessive scouring (erosion) and uplift and helped foster engineering interest in increasing the cross-sectional dimensions of such designs. After years of delay the collapsed Austin Dam was eventually replaced and is now known as the Tom Miller Dam.

■ **Austin Dam**
Across Freeman's Run
Austin, Pa.
Bayless Pulp and
Paper Company
1909–11

As part of its Austin paper mill complex in the north-central part of Pennsylvania, the Bayless Pulp and Paper Company built a 50-foot-high, 540-foot-long, concrete gravity storage dam. The foundation conditions at the site were less than ideal and failed to provide a solid, nonporous base of support for the structure. Despite early indications that the dam could not safely resist the horizontal thrust of water pressure exerted by the reservoir, the company persisted in its plans to use the

structure without making substantive improvements in the design. Leakage under the dam gradually eroded the foundations, and on September 30, 1911, the dam failed by sliding along its base. The collapse is considered a classic case of water penetrating under a gravity dam and then pushing up on the bottom of the structure. This "uplift" reduces the ability of the dam to resist horizontal forces and greatly increases the possibility that it will dislodge from its foundations in a downstream direction. Following the collapse of the Austin Dam, it was never rebuilt. Today, the remains of the concrete structure are still visible from State Route 872.

After the San Francisco earthquake of 1906, the greatest disaster in modern California history resulted from the collapse of the St. Francis Dam on the night of March 12–13, 1928. The 208-foot-high, concrete curved gravity dam was built by the Los Angeles Bureau of Water and Power to store water from the municipally owned Owens Valley Aqueduct. The reservoir behind the dam did not completely fill with water until early March 1928, two years after it was completed. Following reports of excessive leakage under the dam, its designer, William Mulholland, inspected the structure on March 12 and reported it to be perfectly safe. But late that night the dam suddenly gave way and released an enormous amount of water into the San Francisquito Canyon and the Santa Clara River Valley. By the time the deluge finally washed into the Pacific Ocean, approximately 400 persons were dead and millions of dollars of agricultural land lay in ruin. Unstable rock foundations at the site were eventually blamed for the failure. Responsibility also was placed on both Mulholland and a loophole in California's dam safety laws that exempted Los Angeles from any outside supervision over its dam-building activities. As a result of the St. Francis disaster, California passed a new dam safety law that became the most stringent in the United States. City authorities dynamited all remains of the dam in hopes that it would help people forget the tragedy. Sharp-eyed travelers, however, can still see large chunks of concrete in San Francisquito Canyon several miles downstream from the former dam site.

■ **St. Francis Dam**
Across San Francisquito Creek
Saugus, Calif.
William Mulholland
1926–28

St. Francis Dam. Left: Dotted line indicating portion left standing after the collapse. Right: Dam's remains in mid-March 1928.

FURTHER READING

Allen, Richard Saunders. *Covered Bridges of the Middle Atlantic States.* Brattleboro, Vt.: Stephen Greene Press, 1959.

_____. *Covered Bridges of the Northeast.* Brattleboro, Vt.: Stephen Greene Press, 1957.

_____. *Covered Bridges of the South.* Brattleboro, Vt.: Stephen Greene Press, 1970.

American Society of Civil Engineers. *American Wooden Bridges.* Committee on History and Heritage of American Civil Engineering. New York: Author, 1976.

Baker, T. Lindsay. *Building the Lone Star: An Illustrated Guide to Historic Engineering Sites in Texas.* College Station: Texas A&M Press, 1986.

Billington, David P. *The Tower and the Bridge: The New Art of Structural Engineering.* New York: Basic Books, 1983.

Bissell, Charles A. *History and First Annual Report of the Metropolitan Water District of Southern California.* Los Angeles: Haynes Corporation, 1939.

Blake, Nelson M. *Water for the Cities.* Syracuse, N.Y.: Syracuse University Press, 1956.

Bluestone, Daniel M. *Cleveland: An Inventory of Historic Engineering and Industrial Sites.* Washington, D.C.: Historic American Engineering Record, 1978.

Brooklyn Museum. *The Great East River Bridge, 1883–1983.* New York: Abrams, 1983.

Bureau of Reclamation. *Dams and Control Works.* Washington, D.C.: U.S. Government Printing Office, 1929.

Caro, Robert A. *The Path to Power.* New York: Alfred A. Knopf, 1982.

Chamberlin, William P. *National Cooperative Highway Research Program, Synthesis of Highway Practice 101: Historic Bridges — Criteria for Decision Making.* Washington, D.C.: Transportation Research Board, National Research Council, 1983.

Citizen's Committee to Save the Littlerock Dam, Inc. v. Ronald B. Robie, et al.; Superior Court of California, County of Los Angeles, C184 269; April 20, 1977.

Comp, T. Allan, and Donald C. Jackson. *Bridge Truss Types: A Guide to Dating and Identifying.* Technical Leaflet Series, no. 95. Nashville: American Association for State and Local History, 1977.

Condit, Carl. *American Building Art: The Nineteenth Century.* New York: Oxford University Press, 1960.

_____. *American Building Art: The Twentieth Century.* New York: Oxford University Press, 1961.

Darnell, Victor. *Directory of American Bridge Building Companies, 1840–1900.* Washington, D.C.: Society for Industrial Archeology, 1985.

Davis, Arthur Powell. *Irrigation Works Constructed by the United States Government.* New York: John Wiley and Sons, 1917.

Davison, Stanley. *The Leadership of the Reclamation Movement, 1875–1902.* New York: Arno Press, 1979.

Deibler, Dan. *A Survey and Photographic Inventory of Metal Truss Bridges in Virginia, 1865–1932.* 5 vols. Charlottesville: Virginia Highway and Transportation Research Council, 1975–76.

Draper, Joan E. *Chicago Bridges.* Naomi Donson, ed. Chicago: Chicago Department of Public Works, 1984.

Edwards, Llewellyn Nathaniel. *A Record of History and Evolution of Early American Bridges.* Orono: Maine University Press, 1959.

Fraser, Clayton. *Historic Bridges of Colorado.* Denver: Colorado Department of Highways, 1987.

Historic American Engineering Record. *HAER Checklist: 1969–1985.* Washington, D.C.: National Park Service, 1985.

Hopkins, H. J. *A Span of Bridges: An Illustrated History.* New York: Praeger, 1970.

Hughes, Thomas P. *Networks of Power: Electrification in Western Society, 1880–1930.* Baltimore: Johns Hopkins University Press, 1984.

Hyde, Charles K. *The Lower Peninsula of Michigan: An Inventory of Historic Engineering and Industrial Sites.* Washington, D.C.: Historic American Engineering Record, 1976.

———. *The Upper Peninsula of Michigan: An Inventory of Historic and Industrial Sites.* Washington, D.C.: Historic American Engineering Record, 1978.

Jackson, Donald C. "A History of Water in the American West: John S. Eastwood and the 'Ultimate Dam,' 1908–1924," Ph.D. dissertation, University of Pennsylvania, 1986.

———. Nancy C. Shanahan, Elizabeth Sillin and Vincent Marsh. *Saving Historic Bridges.* Information Series, no. 36. Washington, D.C.: National Trust for Historic Preservation, 1984.

Latimer, Margaret, Brooke Hindle and Melvin Kranzberg. *Bridge to the Future: A Centennial Celebration of the Brooklyn Bridge.* New York: New York Academy of Sciences, 1984.

Lee, Lawrence. *Reclaiming the American West: An Historiography and Guide.* Santa Barbara, Calif.: ABC-Clio, 1980.

Leonhardt, Fritz. *Bridges.* Cambridge, Mass.: MIT Press, 1984.

McCullough, David. *The Great Bridge: The Epic Story of the Building of the Brooklyn Bridge.* New York: Simon and Schuster, 1972.

Miller, Howard S. *The Eads Bridge.* Columbia: University of Missouri Press, 1979.

Mock, Elizabeth B. *The Architecture of Bridges.* 1949. Reprint. New York: Arno Press, 1972.

Moeller, Beverly Bowen. *Phil Swing and Boulder Dam.* Berkeley: University of California Press, 1971.

Molloy, Peter M. *The Lower Merrimack River Valley: An Inventory of Historic Engineering and Industrial Sites.* Washington, D.C.: Historic American Engineering Record, 1976.

Myer, Donald B. *Bridges and the City of Washington.* Washington, D.C.: U.S. Commission of Fine Arts, 1974.

Ohio Department of Transportation. *The Ohio Historic Bridge Inventory, Evaluation and Preservation Plan.* Columbus: Author, 1983.

Outland, Charles F. *Man-Made Disaster: The Story of the St. Francis Dam.* Glendale, Calif.: Arthur H. Clark Company, 1963.

Pennsylvania Department of Transportation and Pennsylvania Historical and Museum Commission. *Historic Highway Bridges in Pennsylvania.* Harrisburg: Commonwealth of Pennsylvania, 1986.

Pisani, Donald J. *From the Family Farm to Agribusiness: The Irrigation Crusade in California and the West, 1850–1931.* Berkeley: University of California Press, 1984.

Plowden, David. *Bridges: The Spans of North America.* New York: W. W. Norton, 1974.

Quivik, Fredric L. *Historic Bridges in Montana.* Washington, D.C.: National Park Service, 1982.

Rae, Steven R., Joseph E. King and Donald R. Abbe. *New Mexico Historic Bridge Survey.* Santa Fe: New Mexico State Highway and Transportation Department, 1987.

Reier, Sharon. *The Bridges of New York.* New York: Quadrant Press, 1977.

Robinson, Michael C. *Water for the West: The Bureau of Reclamation, 1902–1977.* Chicago: Public Works Historical Society, 1979.

Roth, Matthew. *Connecticut: An Inventory of Historic Engineering and Industrial Sites.* Washington, D.C.: Society for Industrial Archeology, 1981.

Schuyler, James D. *Reservoirs for Irrigation, Water Power and Domestic Water Supply.* New York: John Wiley and Sons, 1909.

Shank, William H. *Historic Bridges in Pennsylvania.* Rev. ed. York, Pa.: American Canal and Transportation Center, 1980.

Smith, Dwight A., James B. Norman and Pieter T. Dykman. *Historic Highway Bridges of Oregon.* Salem: Oregon Department of Transportation, 1985.

Smith, Karen. *The Magnificent Experiment: Building the Salt River Project, 1870–1917.* Tucson: University of Arizona Press, 1986.

Smith, Norman. *A History of Dams.* Secaucus, N.J.: Citadel Press, 1972.

Spero, Paula A. C. *Metal Truss Bridges in Virginia: 1865–1932.* 4 vols. Charlottesville: Virginia Highway and Transportation Research Council, 1978–82.

Steinman, David Barnard, and Sara Ruth Watson. *Bridges and Their Builders.* 1941. Reprint. New York: Dover Publications, 1957.

Tyrrell, Henry G. *History of Bridge Engineerihg.* Chicago: Author, 1911.

U.S. Army Corps of Engineers, Omaha District. *The Federal Engineer: Damsites to Missile Sites.* Omaha: Author, 1985.

Van der Zee, John. *The Gate.* New York: Simon and Schuster, 1987.

Vogel, Robert M., ed. *Report of the Mohawk-Hudson Area Survey.* Washington, D.C.: National Museum of History and Technology, 1973.

Waddell, J. A. L. *Bridge Engineering.* 2 vols. New York: John Wiley and Sons, 1916.

Wegmann, Edward. *The Design and Construction of Dams.* New York: John Wiley and Sons, 1927.

Welsh, Michael. *A Mission in the Desert.* Albuquerque: Albuquerque District, U.S. Army Corps of Engineers, 1985.

Westbrook, Nicholas, ed. *A Guide to the Industrial Archeology of the Twin Cities.* Minneapolis: Society for Industrial Archeology, 1983.

Willingham, William F. *Water Power in the "Wilderness": The History of Bonneville Lock and Dam.* Portland, Ore.: Portland District, U.S. Army Corps of Engineers, 1987.

Woollett, William. *Hoover Dam: Drawings, Etchings, Lithographs, 1931–1933.* California Architecture and Architects Series. Los Angeles: Hennessey and Ingalls, 1986.

Zuk, William, et al. *Methods of Modifying Historic Bridges for Contemporary Use.* Charlottesville: Virginia Highway and Transportation Research Council, 1980.

■ ■ ■ ■ ■ ■ ■ ■ ■ ■ ■ ■ ■ ■ ■ ■ ■ ■

INFORMATION SOURCES

The following organizations and agencies can provide further information on subjects covered in *Great American Bridges and Dams*. In addition, state historic preservation offices, state historical societies and state transportation and public works departments often have information on the construction and use of bridges and dams.

American Society of
Civil Engineers
Committee on History
and Heritage of
American Civil Engineering
Contact: Herbert Hands
345 East 47th Street
New York, N.Y. 10017

National Park Service,
U.S. Department of
the Interior:

Historic American
Engineering Record
P.O. Box 37127
Washington, D.C. 20013-7127

National Register of
Historic Places
P.O. Box 37127
Washington, D.C. 20013-7127

National Society for
the Preservation of
Covered Bridges
c/o Mrs. Arnold L. Ellsworth
44 Cleveland Avenue
Worcester, Mass. 01603

National Trust for
Historic Preservation
1785 Massachusetts
Avenue, N.W.
Washington, D.C. 20036

Regional Offices:

Northeast Regional Office
45 School Street, 4th Floor
Boston, Mass. 02108

Mid-Atlantic Regional Office
6401 Germantown Avenue
Philadelphia, Pa. 19144

Regional Offices Continued:

Southern Regional Office
456 King Street
Charleston, S.C. 29403

Midwest Regional Office
53 West Jackson Boulevard
Suite 1135
Chicago, Ill. 60604

Mountains/Plains
Regional Office
511 16th Street, Suite 700
Denver, Colo. 80202

Texas/New Mexico
Field Office
500 Main Street, Suite 606
Fort Worth, Tex. 76102

Western Regional Office
One Sutter Street, Suite 707
San Francisco, Calif. 94104

Public Works Historical
Society
1313 East 60th Street
Chicago, Ill. 60637

Society for Industrial
Archeology
c/o Smithsonian Institution
National Museum of
American History
Room 5020
Washington, D.C. 20560

U.S. Army Corps
of Engineers
Office of History
Kingman Building
Fort Belvoir, Va. 22060-5577

■ ■ ■ ■ ■ ■ ■ ■ ■ ■ ■ ■ ■ ■ ■ ■ ■ ■

PHOTOGRAPHIC SOURCES

Abbreviations used refer to the following collections:

CTPC — Curt Teich Postcard Collection, Lake County Museum

DCJ — Donald C. Jackson photographs and postcards from the author's postcard collection

HAER — Historic American Engineering Record, National Park Service

NTHP — National Trust for Historic Preservation

SI — Smithsonian Institution

2 all, DCJ. **7** all, DCJ. **8** Jack E. Boucher, HAER. **10** Jet Lowe, HAER. **14** HAER. **16** all, DCJ. **17** top, DCJ; bottom, HAER. **18** DCJ. **19** all, DCJ. **20** both, DCJ. **21** all, Arnold David Jones, HAER. **22** drawings, Arnold David Jones, HAER; photos, DCJ, except third from top, Arthur C. Huskell, Historic American Buildings Survey. **24** all, Arnold David Jones, HAER. **25** drawings, Arnold David Jones, HAER; photos, DCJ. **26** drawings, Arnold David Jones, HAER; photos, DCJ. **27** drawing, Arnold David Jones, HAER; photos, DCJ. **28** drawings, Arnold David Jones, HAER; photo, DCJ. **29** drawing, Arnold David Jones, HAER; photos, DCJ. **30** both, DCJ. **31** both, DCJ. **32** DCJ. **33** both, DCJ. **34** both, DCJ. **35** both, DCJ. **36** top, DCJ; bottom, Jet Lowe, HAER. **37** both, DCJ. **38–39** both, DCJ. **40** Library of Congress. **42** both, DCJ. **43** both, DCJ. **44** DCJ. **45** top, SI; others, DCJ. **46** both, DCJ. **47** top, Richard K. Anderson; center, SI; bottom, Salt River Project. **48** top, SI; bottom, Richard K. Anderson. **49** top, DCJ; bottom, SI. **50** SI. **51** Water Resources Center Archives, Berkeley, Calif. **52** bottom, DCJ. **52–53** National Museum of American Art, SI. Transfer from the National Park Service, U.S. Department of the Interior. **53** bottom, Library of Congress. **54** Mitch Toll. **55** DCJ. **56** Joe Young, *Indianapolis News*. **57** top, DCJ; bottom, Standard Oil of New Jersey Collection, University of Louisville Photographic Archives. **58** DCJ. **60** DCJ. **63** DCJ. **64** DCJ. **65** top, University of Vermont Historic Preservation Program; bottom, HAER. **66** *Grand Rapids Press*. **69** top, HAER; bottom, NTHP. **72** Salt River Project. **73** top, DCJ; others, SI. **75** all, DCJ. **76** both, DCJ. **79** Janet Walker. **81** Chicago Historical Society.

NEW ENGLAND

82–83 HAER. **84** Matthew Roth, Connecticut Historical Commission/HAER. **85** top, DCJ; bottom, both, Jet Lowe, HAER. **86** Matthew Roth, Connecticut Historical Commission/HAER. **87** top, Matthew Roth, Connecticut Historical Commission/HAER; bottom, HAER. **88** top, HAER; others, Maine Historic

Preservation Commission.   **89** top, DCJ; others, Jet Lowe, HAER.   **90** top, HAER; others, Maine Historic Preservation Commission.   **91** top, HAER; bottom, DCJ.   **92** left, DCJ; right, HAER.   **93** top, DCJ; bottom, Massachusetts Historical Commission.   **94** top, DCJ; bottom, HAER.   **95** HAER.   **96** Jack Maley, Metropolitan District Commission.   **97** New Hampshire Division of Historical Resources.   **98** HAER.   **99** top, B. Clouette, New Hampshire Division of Historical Resources; bottom, David Ruell, New Hampshire Division of Historical Resources.   **100** David R. Proper, New Hampshire Division of Historical ResourcesM.   **101** CTPC.   **102** DCJ.   **103** DCJ. **104** both, Vermont Division for Historic Preservation. **105** both, Vermont Division for Historic Preservation. **106** both, Vermont Division for Historic Preservation. **107** top, American Society of Civil Engineers; bottom, HAER.

MID-ATLANTIC

108–09 Jet Lowe, HAER.   **110** SI.   **111** all, HAER. **112** both, J. Alexander, U.S. Commission of Fine Arts. **113** DCJ.   **114** top left, Leet Brothers, Washington, D.C.; others, U.S. Commission of Fine Arts.   **115** top, DCJ; bottom, SI.   **116** top, U.S. Commission of Fine Arts; bottom, DCJ.   **117** DCJ.   **118** top, DCJ; bottom, HAER.   **119** top left, CTPC; top right, DCJ; bottom, HAER.   **120** both, HAER.   **121** M. E. Warren.   **122** HAER.   **123** left, DCJ; right, Terry Karschner, Office of New Jersey Heritage, New Jersey Department of Environmental Protection.   **124** top, HAER; bottom, DCJ. **125** both, DCJ.   **126** DCJ.   **127** top and center, DCJ; bottom, HAER.   **128** HAER.   **129** HAER.   **130** Library of Congress.   **131** Jet Lowe, HAER.   **132** both, Jet Lowe, HAER.   **133** both, William P. Chamberlin. **134** both, Jet Lowe, HAER.   **135** top, HAER; bottom, DCJ.   **136** DCJ.   **137** both, DCJ.   **138** DCJ.   **139** DCJ.   **140** top right, HAER; others, DCJ.   **141** New York City Department of Environmental Protection. **142** top, Marion Bernstein, New York City Department of Environmental Protection; bottom left, HAER; bottom right, DCJ.   **143** top, Pennsylvania Historical and Museum Commission; center, DCJ; bottom, CTPC. **144** left, Pennsylvania Historical and Museum Commission; right, CTPC.   **145** both, DCJ.   **146** SI.   **147** all, HAER.   **149** all, DCJ.   **150** Pennsylvania Historical and Museum Commission.   **151** Pennsylvania Historical and Museum Commission.   **152** top, both, Jack E. Boucher, HAER; bottom, Pennsylvania Historical and Museum Commission.   **153** top, Frederick R. Love, HAER; bottom, West Virginia Department of Culture and History/HAER.   **154** both, Jet Lowe, HAER. **155** Tony P. Wrenn.   **156** Michael Keller, West Virginia Department of Culture and History.   **157** both, West Virginia Department of Culture and History; top, R. P. Davis; bottom, Rodney S. Collins.   **158** HAER.   **159** HAER.

SOUTH

160–61 Library of Congress.   **163** top, Alabama Historical Commission; center, Library of Congress; bottom, DCJ.   **164** top, Library of Congress; bottom, Alabama Historical Commission.   **165** both, Arkansas Historic Preservation Program; bottom, Michael Swanda.   **166** Arkansas Historic Preservation Program.   **167** all, Arkansas Historic Preservation Program; top and center, Jeff Holder; bottom, Jaci Carfagno.   **168** both, HAER.   **169** Historic St. Augustine Preservation Board.   **170** top, HAER; bottom, Georgia Department of Natural Resources.   **171** both, Georgia Power Company.   **172** bottom left, CTPC; others, Georgia Power Company.   **173** David Plowden.   **174** DCJ.   **175** HAER.   **176** both, The Historic New Orleans Collection.   **177** all, Mississippi Department of Archives and History.   **178** Mississippi Department of Archives and History.   **180** left, North Carolina Division of Archives and History, Department of Cultural Resources; right, DCJ.   **181** Library of Congress.   **183** top, HAER; others, DCJ.   **184** both, HAER.   **185** both, DCJ.   **186** DCJ.   **187** both, Clayton B. Fraser.   **188** top, both, Library of Congress; bottom, DCJ.   **189** DCJ.   **190** Virginia Division of Tourism.   **191** all, HAER; drawing, Charles King.   **192** all, Virginia Department of Conservation and Historic Resources.   **193** top, both, Virginia Department of Conservation and Historic Resources; bottom, DCJ.

MIDWEST

194–95 Jet Lowe, HAER.   **196** CTPC.   **197** both, Illinois Historic Preservation Agency.   **198** top right, DCJ; others, Illinois Historic Preservation Agency.   **199** M. Turner, HAER.   **200** James L. Cooper.   **201 James L. Cooper.**   **202** James L. Cooper.   **203** James L. Cooper.   **204** DCJ.   **205** both, State Historical Society of Iowa.   **206** left, DCJ; right, Kansas State Historical Society.   **207** both, Larry Jochims, Kansas State Historical Society.   **208** Larry Jochims, Kansas State Historical Society.   **209** *Allegan County News and Gazette.*   **210** Balthazar Korab Ltd.   **211** top, Jet Lowe, HAER; bottom, HAER.   **212** Balthazar Korab Ltd.   **213** DCJ.   **215** top, DCJ; bottom, U.S. Army Corps of Engineers.   **216** U.S. Army Corps of Engineers.   **217** top, Clayton B. Fraser; bottom, DCJ.   **218** both, DCJ.   **219** Bonnie Wright, Missouri Department of Natural Resources.   **220** U.S. Army Corps of Engineers.,   **221** all, HAER.   **222** left, both, Gerald Lee Gilleard, Missouri Department of Natural Resources; right, Clayton B. Fraser.   **223** Nebraska State Historical Society.   **224** North Dakota State Historical Society.   **225** Christopher Duckworth and William Keener, Ohio Historical Society.   **226** Library of Congress.   **227** top, SI; others, HAER.   **228** top, SI; bottom, David Simmons, Ohio Historical Society.   **229** all, Jet Lowe, HAER.   **230** top, HAER; others, Jet Lowe, HAER.

231 David L. Taylor, Ohio Historical Society.   232 top, David Simmons, Ohio Historical Society; bottom, Ohio Historical Society.   233 Ohio Historical Society.   234 DCJ.   235 both, DCJ.   236 top, U.S. Army Corps of Engineers; bottom, South Dakota State Historical Society.   237 South Dakota State Historical Preservation Center.   238 both, State Historical Society of Wisconsin.   239 top, Donald N. Anderson; both, State Historical Society of Wisconsin.

SOUTHWEST

240–41 Salt River Project.   242 top, Clayton B. Fraser; bottom, Salt River Project.   243 SI.   244 top, both, DCJ; bottom, Clayton B. Fraser.   245 Arizona Office of Tourism.   246 top, Clayton B. Fraser; bottom, both, Salt River Project.   248 left, Salt River Project; right, DCJ.   249 top, Clayton B. Fraser; bottom, DCJ.   250 U.S. Army Corps of Engineers.   251 top, New Mexico Historic Preservation Division; others, DCJ.   252 New Mexico Historic Preservation Division.   253 New Mexico Historic Preservation Division.   254 both, DCJ.   255 both, Oklahoma Historical Society, Museum of the Western Prairie.   256 top, Lower Colorado River Authority; others, DCJ.   257 top, DCJ; bottom, Rosenberg Library, Galveston, Tex.   258 top, Lower Colorado River Authority; bottom, DCJ.   259 CTPC.

WEST

260–61 Library of Congress.   262 both, Jet Lowe, HAER.   263 top, DCJ; bottom, SI.   265 top, DCJ; center, CalTrans; bottom, Water Resources Center Archives, Berkeley, Calif.   266 Bureau of Reclamation.   267 top, Water Resources Center Archives, Berkeley, Calif.; bottom, CalTrans.   268 top, Library of Congress; bottom, Jet Lowe, HAER.   269 HAER.   270 top, HAER; others, Jet Lowe, HAER.   271 DCJ.   272 DCJ.   273 bottom right, Jet Lowe, HAER; others, DCJ.   274 both, HAER.   275 both, DCJ.   276 CalTrans.   277 top, DCJ; bottom, HAER.   278 DCJ.   279 all, Jet Lowe, HAER.   280 top, DCJ; bottom, San Francisco Public Utilities Commission.   281 Library of Congress.   282 CalTrans.   283 both, Clayton B. Fraser.   284 top, Clayton B. Fraser; bottom, Colorado Historical Society.   285 both, Clayton B. Fraser.   286 both, Guam Department of Parks and Recreation.   287 both, DCJ.   288 top, *Engineering,* November 9, 1888, and DCJ; others, Idaho State Historical Society.   289 top, Idaho Department of Natural Resources; bottom, Idaho State Historical Society.   290 bottom left, DCJ; others, Idaho State Historical Society.   291 both, HAER.   292 all, HAER.   293 U.S. Army Corps of Engineers.   294 top, both, HAER; bottom, both, Jet Lowe, HAER.   295 Montana Power Company.   296 top left and center left, DCJ; others, Library of Congress.   297 both, DCJ.   298 Nevada Historical Society.   **299** both, Nevada Historical Society.   **300** Nevada Historical Society.   **301** all, U.S. Army Corps of Engineers.   **302** both, DCJ.   **303** both, DCJ.   **304** DCJ.   **305** left, Library of Congress; right, DCJ.   **307** both, Utah State Historical Society.   **308** all, HAER.   . **309** top, HAER; bottom, Utah State Historical Society. **310** top, both, Utah State Historical Society; bottom, Jack E. Boucher, HAER.   **311** all, DCJ.   **313** both, DCJ. **314** top, DCJ; bottom, Jet Lowe, HAER.   **315** top, DCJ; bottom, Jet Lowe, HAER.   **316** top, Jet Lowe, HAER; bottom, Wyoming State Archives, Museums and Historical Department.   **317** left, HAER; right, Mark Junge, Wyoming State Archives, Museums and Historical Department.   **318** HAER.   **319** top, both, HAER; bottom, Clayton B. Fraser.

320 SI.   **322** NTHP.   **323** top to bottom, Library of Congress; SI; Jack Boucher, HAER; DCJ.   **324** Library of Congress.   **325** top, HAER; bottom, United Press International.   **326** DCJ.   **328** both, SI.   **329** SI. **330** DCJ.   **331** both, SI.   **360** DCJ.

INDEX

AUTHOR

Donald C. Jackson of Philadelphia is a historian of technology involved in a wide range of projects as a consultant, lecturer and principal investigator. Most recently, he taught courses on the history of technology at the University of Pennsylvania. From 1975 to 1985 Jackson served as an engineering historian with the National Park Service's Historic American Engineering Record, Washington, D.C., where he supervised inventory and recording projects for historic engineering sites, prepared historical reports and lectured extensively. He held a research fellowship at the Smithsonian Institution's National Museum of American History from 1985 to 1986. Jackson holds a B.S. in civil engineering from Swarthmore College and a Ph.D. in American civilization from the University of Pennsylvania, where he wrote his dissertation on the history of the multiple-arch dam. He is also an author of the National Trust's publication *Saving Historic Bridges.*

David McCullough is author of the acclaimed book on the building of the Brooklyn Bridge, *The Great Bridge,* as well as other books including *The Johnstown Flood,* *The Path Between the Seas* and *Mornings on Horseback.* He is also host of the PBS series "Smithsonian World," for which he received an Emmy, and "The American Experience," a new series on American history. A native of Pittsburgh now living in West Tisbury, Mass., McCullough lectures widely and is a contributing editor for a number of periodicals. He has received numerous awards for his work, including honorary degrees in engineering and the humanities.

Turn-of-the-century bridge lovers enjoying a walk across a swinging bridge. Be careful out there!